Immobilised cells and enzymes

a practical approach

TITLES PUBLISHED IN
—— THE ——
PRACTICAL APPROACH
—— SERIES ——

Immobilised cells and enzymes

a practical approach

Edited by
J Woodward

Oak Ridge National Laboratory, Oak Ridge, TN 37831, USA

Mylar

Sep

Chem

◇ IRL PRESS
Oxford · Washington DC

7274 - 7936

IRL Press Limited
P.O. Box 1,
Eynsham,
Oxford OX8 1JJ,
England

CHEMISTRY

©1985 IRL Press Limited

British Library Cataloguing in Publication Data

Immobilised cells and enzymes: a practical approach.
 —(Practical approach series)

 1. Immobilised cells — Laboratory manuals
 2. Immobilised enzymes — Laboratory manuals
 I. Woodward, Jonathan   II. Series
       574.87'028   QH585.5.I45

ISBN 0-947946-21-7

Cover Photograph: Bioreactors for the energy-conserving anaerobic treatment of wastewaters. Fixed-films of bacterial cells are attached to 2.5-cm inert ceramic Raschig rings to confine them within the continuous-flow system. The films formed after inoculating the bioreactors with sludge from a common anaerobic digester from a conventional municipal wastewater treatment plant. Kindly supplied by the Oak Ridge National Laboratory, Oak Ridge, Tennessee, U.S.A.

Printed in England by Information Printing, Oxford.

Preface

The compilation of the chapters in this book has been done with the intention of introducing and instructing students and readers into the art of cell and enzyme immobilisation by describing in detailed, recipé-type fashion, the apparatus, materials, and methods used. While several chapters are general in nature, others address more specific applications of immobilised cells and enyzmes. There is much current interest in using biotechnology for the production of fuels and chemicals (including pharmaceuticals and hormones), and in this regard the use of immobilised cells and enzymes in biotechnological processes has tremendous potential. It is hoped, therefore, that this practical book will serve the interests of industry, academia, and governments alike.

I am extremely indebted to the authors who, through their contributions, have, in my opinion, made this a valuable handbook. I would also like to thank Dr. Alan Wiseman of Surrey University who is responsible for my interest in immobilised cells and enzymes. Finally, I am also grateful to Ms. Debbie Weaver of Oak Ridge National Laboratory for her excellent secretarial service.

Jonathan Woodward

Editorial note

The editorial duties associated with the compilation of this book were supported in part by the Office of Basic Energy Sciences, U.S. Department of Energy, under Contract No. DE-AC05-840R21400 with Martin Marietta Energy Systems, Inc.

Contributors

S.P.Bidey
Department of Molecular Endocrinology, The Middlesex Hospital Medical School, Mortimer Street, London W1N 8AA, UK

P.Brodelius
Institut für Biotechnologie, ETH-Hönggerberg, CH-8093 Zürich, Switzerland

J.M.S.Cabral
Laboratório de Engenharia Bioquimica, Instituto Superior Técnico, Universidade Técnica de Lisboa, 1000 Lisbon, Portugal

M.Coughlan
Department of Biochemistry, University College, Galway, Eire

G.de Olivera Neto
Instituto de Quimica, Cidade Universitaria, Universidade de Sao Paulo, Sau Paulo, Brazil

G.G.Guilbault
Department of Chemistry, University of New Orleans, New Orleans, LA 70148, USA

I.J.Higgins
Joint Director, Cranfield Biotechnology Centre and Leicester Biocentre, Cranfield Institute of Technology, Cranfield, Bedford MK43 0AL, and University Road, Leicester LE1 7RH, UK

J.F.Kennedy
Department of Chemistry, University of Birmingham, P.O.Box 363, Birmingham B15 2TT, UK

M.P.J.Kierstan
Kerry Co-op Creameries Ltd., R & D Centre, Ardfert, Co. Kerry, Eire

H.E.Klei
Department of Chemical Engineering, University of Connecticut, Storrs, CT 06268, USA

K.A.Koshcheyenko
Institute of Biochemistry and Physiology of Microorganisms, USSR Academy of Sciences, Pushchino, Moscow Region 142292, USSR

D.Shim
Department of Chemical Engineering, University of Connecticut, Storrs, CT 06268, USA

G.V.Sukhodolskaya
Institute of Biochemsitry and Physiology of Microorganisms, USSR Academy of Sciences, Pushchino, Moscow Region 142292, USSR

D.W.Sundstrom
Department of Chemical Engineering, University of Connecticut, Storrs, CT 06268, USA

L.B.Wingard,Jr.
Department of Pharmacology, University of Pittsburgh, Pittsburgh, PA 15261, USA

J.Woodward
Chemical Technology Division, Oak Ridge National Laboratory, P.O.Box X, Oak Ridge, TN 37831, USA

Contents

Abbreviations

BSA	bovine serum albumin
cAMP	cyclic AMP
CAS	Concanavalin A-Sepharose
CM	carboxymethyl
CNBr	cyanogen bromide
Con A	Concanavalin A
DEAE	diethylaminoethyl
DMSO	dimethylsulphoxide
EDTA	ethylenediamine tetracetic acid
ELISA	enzyme-linked immunosorbent assay
FAD	flavin-adenine dinucleotide
FDA	fluorescein diacetate
FSH	follicle-stimulating hormone
hCG	human chorionic gonadotropin
Hepes	N-2-hydroxyethylpiperazine-N'-2-ethanesulphonic acid
IC	immobilised cells
LH	luteinising hormone
MBA	N,N'-methylenebisacrylamide
MIX	3-isobutyl-1-methylxanthine
NAD(P)	nicotine-adenine dinucleotide (phosphate)
PEG	polyethylene glycol
PEI	polyethyleneimine
PVA	polyvinyl alcohol
PVP	polyvinylpyrrolidone
RRDE	rotating ring-disc electrode
s.c.e.	saturated calomel electrode
TEMED	N,N,N',N'-tetramethylenediamine
TSH	thyroid-stimulating hormone

Introduction

I.J. HIGGINS

We are rapidly discovering more and more ways of exploiting the unique catalytic and recognition properties of biological systems in disparate areas of human activity, from medicine to the military; from food processing to microelectronic sensing devices. In many cases the development of practical processes or devices incorporating biological elements is critically dependent upon devising appropriate procedures for the retention and stabilisation of the biological component, be it a large molecule or small particle. In many cases therefore, a method of immobilisation is required.

Although immobilisation methods have been exploited for many years in biologically based products, for example, enzyme test strips, the technology has developed rapidly over recent years. Not only has the range of approaches to retaining the biological element expanded significantly but immobilisation is becoming increasingly important for modifying the behaviour of that element. For example, kinetics, specificity and stability can be altered to design a more appropriate system for commercialisation. Further, the technology is becoming increasingly predictive; in other words, more of a science than an art.

There is no doubt that immobilisation methodology will play an increasingly important role in biotechnology and indeed its development represents the rate-limiting step in some areas. Notwithstanding the foregoing, there is much that we do not understand about these processes. For example, recent collaborative work between the Cranfield Biotechnology Centre and the University of Oxford has led to the development of new types of amperometric biosensors in which immobilised oxidases make intimate electronic contact with modified carbon conductors. The immobilisation procedure, together with the chemical modification of the conductor, results in a dramatic (and extremely valuable) change in the kinetics of the enzymes and to electron flow to the conductor instead of to the natural electron acceptor, molecular oxygen. The reason for the phenomenon is not fully understood at the molecular level.

Research into the science of cell and enzyme immobilisation is currently extremely important, will continue to be so for the foreseeable future and will impinge increasingly on developments in enzyme engineering and modification of cells and microorganisms.

This volume offers a 'state of the art' account of the current position in immobilisation technology, with particular emphasis on the practical aspects. It covers a wide field and has been compiled by recognised leaders in the subject. It will be invaluable both to researchers and those requiring detailed information about the practicalities of this important subject.

CHAPTER 1

Immobilised Enzymes: Adsorption and Covalent Coupling

JONATHAN WOODWARD

1. INTRODUCTION

The adsorption of an enzyme onto an insoluble support is the simplest method of enzyme immobilisation. Thus, it is a useful technique with which to initiate a student into the area of immobilised enzymes. The procedure consists of mixing together the enzyme and support material under appropriate conditions and, following a period of incubation, separating the insoluble material from the soluble material by centrifugation or filtration. The major disadvantage of this method is that the enzyme is not firmly bound to the support. For example, the adsorption of enzymes onto an insoluble matrix such as DEAE-Sephadex is mainly due to multiple salt-linkages. Changes in experimental conditions such as pH, ionic strength, temperature and type of solvent can cause desorption of the enzyme from the support as they affect such linkages. It should also be mentioned that the substrate itself can also cause desorption of its enzyme from the support. Besides salt-linkages, other weak binding forces (e.g., hydrogen bonds, Van der Waals forces) are also involved in the adsorption of an enzyme to the support material. Examples of materials to which enzymes have been adsorbed are given in *Table 1*. Ideally, the immobilisation of an enzyme should result in no loss of catalytic activity. This goal can be achieved, for example, when β-glucosidase is immobilised by adsorption onto Concanavalin A-Sepharose (CAS). Note the discussion in reference 1. Generally speaking, adsorption is a mild method of immobilisation and often has little effect on catalytic activity.

Covalent coupling for the immobilisation of enzymes is based upon the formation of a covalent bond between the enzyme molecules and support material. It is important that the amino acids essential to the catalytic activity of the enzyme are not involved in the covalent linkage to the support. This may be difficult to achieve, and enzymes immobilised in this fashion generally lose activity upon immobilisation. This problem may be prevented if the enzyme is immobilised in the presence of its substrate — a step in the procedure which would tend to have a protective effect on the catalytic site of the enzyme during immobilisation. Typical water-insoluble support materials used for the covalent attachment of enzymes are shown in *Table 2*. Prior to the covalent attachment of an enzyme onto the support, the latter must be activated. Once activated, the support can then react with particular groups on the enzyme. Included are the α- and ϵ-groups of lysine, tyrosine, histidine, arginine and cysteine residues. Details of the covalent

Table 1. Materials Used for the Adsorption of Enzymes.

Alumina
Bentonite
Calcium carbonate
Calcium phosphate gel
Carbon
Cellulose
Clay
Collagen
Concanavalin A-Sepharose
Glass, porous
Hydroxyapatite
Ion-exchange resins
Kaolin
Phenolic polymers
Silica gel

Table 2. Materials Used for the Covalent Attachment of Enzymes.

Agarose (Sepharose)
Cellulose
Dextran (Sephadex)
Glass
Polyacrylamide co-polymers
Polyaminostyrene

coupling of an enzyme to a support are given later in this chapter.

It is not the aim of this chapter to review all the methods of enzyme immobilisation by adsorption and covalent coupling. In this regard, the reader is referred to fuller accounts on these subjects such as those by Zaborsky (2), Barker and Kay (3) and Goldstein and Mannecke (4), especially for a theoretical approach. For a comprehensive treatment of immobilised enzyme methodology, the reader is also referred to the volume of *Methods in Enzymology* dealing solely in this area (5).

The aim of this chapter is to exemplify the adsorption and covalent coupling of enzymes to a support material by considering the specific examples, β-fructo-furanosidase (invertase) and β-glucosidase (cellobiase). Procedures used in their immobilisation, activity measurement and property determinations will be described. Such techniques will be applicable to most enzymes, but methods finally chosen will be dictated by the characteristics of the enzyme in question.

The enzymes β-fructofuranosidase (invertase) and β-glucosidase (cellobiase) are particularly suitable for initiating a student into enzyme immobilisation, primarily because they are cheap and readily available. They are both industrially produced, and samples are usually donated free for research purposes. For example, commercial yeast invertase (β-fructofuranosidase) concentrate can be obtained from Honeywill and Stein Co., Ltd., Wallington, Surrey, UK. Cellobiase 250 L is a concentrated β-glucosidase preparation that can be obtained from Novo Laboratories, Inc., Wilton, Connecticut, USA.

2. ADSORPTION BY IONIC BINDING TO DEAE- AND CM-SEPHADEX

2.1 Preparation of the Support Material

One of the properties of Sephadex ion exchangers is that they swell when placed in an aqueous solvent. The degree of swelling is dependent upon the ionic strength of the swelling medium and is greater when the ionic strength is low and *vice versa*. Prior to the adsorption of the enzyme onto the support, take a quantity of dry Sephadex material and swell it in the appropriate buffer to which the enzyme will adsorb. The choice of buffer will depend on the enzyme to be immobilised. For example, baker's yeast invertase will adsorb onto the cationic exchanger DEAE-Sephadex A-50 swollen or equilibrated in 10 mM sodium phosphate buffer, pH 7.0 (6). Take 0.5 g dry weight of DEAE-Sephadex A-50 and add 100 ml of 10 mM sodium phosphate buffer, pH 7.0, and leave at room temperature for 3 days. After this time interval, the support material will be fully swollen. The swelling can be achieved in a few hours if the temperature is increased to 90°C. For immobilisation onto the anionic exchanger CM-Sephadex C-50, follow the same procedure except use 10 mM sodium acetate buffer, pH 3.6, as the equilibrating buffer. Note that a buffer of low ionic strength must be used otherwise adsorption of enzyme to the support will be impaired due to preferential adsorption of the buffer counterion over the protein ion.

2.2 Adsorption of Invertase Activity

2.2.1 *Determination of Free or Soluble Invertase Activity*

It is important to measure the amount of enzyme that is immobilised onto the support. This measurement is usually done by determining the activity of the enzyme that is contacted with the support and that which remains in solution after the support material is separated from it following the immobilisation procedure. The difference between them gives the theoretical or maximum amount of immobilised activity. Measure invertase activity by taking a suitably diluted 0.1 ml aliquot of the enzyme and incubating it with 0.4 ml sucrose solution (50% w/v) and 1.5 ml of 0.1 M sodium acetate buffer, pH 4.7, at 25°C. After suitable time intervals, remove a 0.1 ml aliquot of the reaction mixture and measure its glucose content using one of the glucose assay reagent kits that are commercially available. Note that both the diluted enzyme and sucrose solutions are made up in the buffer solution in which the reaction is being carried out. One unit of activity is defined as the amount of enzyme which will hydrolyse 1 μmol of sucrose per minute under the conditions of the assay. The commercial invertase described in this chapter (obtained from Honeywill and Stein) was found to have an activity of 4000 units/ml.

Table 3. Preparation of Sephadex-invertase Complex.

1.	Add 0.1 ml of commercial invertase (~400 units) to 10 ml of each equilibrated Sephadex (= 50 mg dry weight of support material).
2.	Mix by end-to-end rotation or by very gentle stirring for 30 min at room temperature.
3.	Centrifuge the Sephadex-invertase complexes using a bench-top centrifuge, pour off the supernatant, and resuspend the complexes in their respective equilibrating buffers. Repeat the centrifugation and resuspension steps several times until no activity can be detected in the supernatant. This ensures that any non-adsorbed enzyme is removed.
4.	Finally, resuspend the complexes in their equilibrating buffers (10 ml total volume).

2.2.2 *Coupling of the Enzyme to the Support*

The preparation of the Sephadex-invertase complexes is described in *Table 3* and refers to the mixing- or shaking-bath process. This technique is commonly employed for laboratory preparations.

2.2.3 *Activity of Adsorbed Invertase and Stability of the Enzyme:Support Bond*

The activity of the adsorbed enzyme is measured in exactly the same way as for the soluble or free enzyme, but 0.1 ml of the invertase-Sephadex complex is used in the assay mixture. The Sephadex particles are small enough to be pipetted into the standard-type glass pipettes and plastic pipette tips. It will be found, however, that the particles have a tendency to adhere to the sides of the pipette or pipette tip. Although this is difficult to overcome, complete removal of most particles can be achieved by repeated washing of the pipette or pipette tip with the assay mixture.

Since the optimum pH for yeast invertase is 4.7, it is important to determine that the adsorbed enzyme (at pH 7.0 or 3.6) does not desorb from the support under assay conditions. In other words, the enzyme:support bond is stable. Indeed, the adsorption of a protein to ion-exchange material is known to be sensitive to changes in pH.

The amount of activity adsorbed to DEAE- and CM-Sephadex (measured at the pH for optimum activity) is shown in *Table 4*. These data assume that the enzyme is firmly (stably) bound to the support under the assay conditions and that actual immobilised enzyme activity is being measured. This assumption is true for the enzyme bound to DEAE-Sephadex but *not* for CM-Sephadex (*Table 5*). The higher retention of activity seen when invertase is adsorbed on CM-Sephadex (*Table 4*, column 6) may be due to the fact that about half of the enzyme activity desorbs from this support at pH 4.7, and, therefore, the total activity measured is not truly immobilised enzyme activity.

The true properties of an adsorbed enzyme, including its activity, can only be measured when it has been determined that desorption of the enzyme does not occur under conditions to which the enzyme is subjected (e.g., pH, temperature, ionic strength, substrate and product concentrations). It has been established that invertase adsorbed on the DEAE-Sephadex is firmly bound in 10 mM sodium acetate buffer, pH 4.7, and under these conditions the activity of the adsorbed

Table 4. Preparation of Invertase Adsorbed onto DEAE- and CM-Sephadex.

Support	Invertase added (units)	Invertase in washings (units)	Bound invertase (A) maximum (units)	Activity of complex (B) (units)	B/A (%)
DEAE-Sephadex	400	0	400	164	41
CM-Sephadex	400	100	300	210	70

Table 5. Elution of Invertase Activity from DEAE-Sephadex and CM-Sephadex by Buffers at Various pH Values[a].

Buffer[b] pH	Eluted activity (units)	
	DEAE-Sephadex	CM-Sephadex
2.5	13	0
3.6	0	0
4.0	0	0
4.7	0	16
5.6	0	21
6.0	0	19
7.0	0	22

[a]13 units of DEAE-Sephadex-invertase and 34 units of CM-Sephadex-invertase complexes were shaken for 5 min at room temperature (18°C) in 1.9 ml of buffer, and the activity of the eluant was measured.
[b]Composition of buffer: 10 mM glycine-HCl, pH 2.5; 10 mM sodium acetate, pH 3.6, 4.0, 4.7, 5.6; 10 mM sodium phosphate, pH 6.0, 7.0.

enzyme is also known. Typically, it is desirable to know the following properties of the adsorbed enzyme: pH optimum, pH stability, temperature optimum, temperature stability and K_m. As an example, let us determine whether the enzyme is firmly bound in the presence of its substrate.

(i) Take 0.5 ml of adsorbed enzyme (8.2 units, see *Table 3*) and add to 9.5 ml of 10 mM sodium acetate buffer pH 4.7 containing a known concentration of sucrose.

(ii) Use between 0.05 and 0.3 M (K_m of yeast invertase is ~30 mM) and incubate for various times at room temperature (the latter varies depending on the geographical location).

(iii) After a specified time, centrifuge the adsorbed enzyme-sucrose mixture, remove the supernatant (taking care not to lose any of the adsorbed enzyme material), wash the latter several times with buffer to remove the sucrose and finally resuspend it in buffer. The adsorbed enzyme activity can then be determined.

The method described here can be applied generally to determine under what conditions desorption of the enzyme occurs.

3. ADSORPTION OF ENZYMES ONTO AFFINITY MATERIALS

3.1 Affinity-type Materials

These materials are adsorbents that have affinity for a group of related substances (*Table 6*). They can be purchased commercially in immobilised form being covalently attached to different types of Sepharose, which is a bead-formed agarose gel. For example, Concanavalin A-Sepharose is Concanavalin A (Con A) coupled to Sepharose 4B by the cyanogen bromide method (see Section 4). Con A is a lectin (a protein having the ability to react reversibly with specific sugar residues) that binds the mannose, glucose and sterically related residues. Con A-Sepharose is thus useful for the purification of glycoproteins containing these sugars. A glycoenzyme can also be immobilised by adsorption to Con A-Sepharose, providing it possesses mannose or glucose residues. The enzymes invertase and cellobiase are good candidates for adsorbing to Con A-Sepharose since both are glycoproteins (7,8).

3.2 Preparation of and Coupling of Cellobiase to the Support Material

Con A-Sepharose is sold as a slurry in 0.1 M acetate at pH 6.0, containing 1 M NaCl and 10^{-3} M $MnCl_2$, $MgCl_2$, $CaCl_2$, and 0.01% merthiolate as a preservative. Note that the binding of glycoproteins to Con A is dependent upon the presence of both Mn^{2+} and Ca^{2+}. However, the glycoprotein-Con A complex is stable in the absence of exogenously added Mn^{2+} and Ca^{2+} at neutral pH. At pH values below 5, the mixture should contain those cations to preserve the binding activity of Con A. The procedures used for the immobilisation to cellobiase to Con A-Sepharose are given in *Table 7*. The cellobiase, which is a very concentrated and viscous liquid, is that provided by Novo Enzymes.

Table 6. Affinity-type Materials and their Specificity[a].

Protein A-Sepharose CL-4B	F_c region of IgG and related molecules
Con A-Sepharose	Terminal α-D-glucopyranosyl, α-D-mannopyranosyl or sterically similar residues
Lentil lectin-Sepharose 4B	Similar to that of Con A-Sepharose but lower binding affinity for simple sugars
Wheat germ lectin-Sepharose 6MB	N-acetyl-D-glucosamine
Poly(U)-Sepharose 4B	Nucleic acids, especially mRNA, which contain poly(A) sequences; poly(U)-binding proteins
Poly(A)-Sepharose 4B	Nucleic acids and oligonucleotides which contain poly(U) sequences; RNA-specific proteins
Lysine-Sepharose 4B	Plasminogen; ribosomal RNA
Blue Sepharose CL-6B	Broad range of enzymes which have nucleotide co-factors; serum albumin; etc.
5'-AMP-Sepharose 4B	Enzymes which have NAD^+ as co-factor and ATP-dependent kinases
2',5'-ADP-Sepharose 4B	Enzymes which had $NADP^+$ as co-factor

[a]Data obtained from *Affinity Chromatography, Principles and Methods,* Pharmacia Fine Chemicals.

Table 7. Immobilisation of Cellobiase on Con A-Sepharose.

1.	Take the required volume of uniformly suspended slurry and wash thoroughly with 10 mM soldium phosphate buffer, pH 7.4, containing 0.5 M NaCl and 1 mM CaCl$_2$, and resuspend in this buffer to the original volume.
2.	Stir 5.0 ml of cellobiase (diluted by a factor of 100 with 10 mM sodium phosphate buffer, pH 7.4) with 5.0 ml of Con A-Sepharose (CAS) slurry at room temperature (~20°C) for 30 min.
3.	Filter the CAS-enzyme complex through a medium-sized porous glass filter and then thoroughly wash the CAS-enzyme complex on the filter with 10 ml portions of 10 mM sodium acetate buffer, pH 4.8, containing 1 mM CaCl$_2$ and 1 mM MnCl$_2$ until no enzyme activity is measurable in the washings.
4.	Resuspend the CAS-enzyme complex in 10 ml of the buffer (pH 4.8, containing Ca^{2+} and Mn^{2+}).

3.2.1 *Determination of Free or Soluble Cellobiase Activity*

Take 0.1 ml of the concentrated enzyme (Novo) diluted (1:200) in 10 mM sodium acetate buffer, pH 5.0, and add to 4.9 ml of the same buffer containing 10 mM cellobiose at 30°C. Monitor the formation of glucose, which is linear for up to 30 min. (One unit of cellobiase is defined as the amount of enzyme needed to produce 1.0 μmol glucose/min at 30°C.) Under these conditions the activity of cellobiase is 3.6 units per ml of dilute enzyme.

3.2.2 *Activity of Adsorbed Cellobiase and Stability of the Enzyme: Support Bond*

Measure the activity of the CAS-cellobiase complex in exactly the same way as for the free enzyme except that 0.1 ml of the adsorbed enzyme slurry is added to the assay mixture. Sepharose particles also adhere to the surface of glass pipettes and plastic pipette tips; hence, the assay mixture must be used to wash the pipette or pipette tip in order that all the particles are added to the assay mixture.

The adsorption of a glycoprotein such as cellobiase to an affinity material such as Con A depends upon the interaction between the sugar residues covalently attached to the enzyme and the Con A molecule. It is thus important to determine that the substrate cellobiose and product glucose do not cause desorption of the enzyme from the support since they are both sugars and because glucose binds to Con A.

(i) Take 0.5 ml of the CAS-cellobiase slurry and add it to 4.5 ml of 10 mM sodium acetate buffer, pH 4.8, containing 1 mM Ca^{2+} and 1 mM Mn^{2+} and either cellobiose or glucose to give a final concentration of 0.1 or 0.5 M, respectively.

(ii) Incubate the mixture at either room temperature or 50°C for 1 h.

(iii) Centrifuge the mixture, carefully remove the supernatant to prevent the loss of any of the CAS-cellobiase slurry, and wash the slurry thoroughly in buffer, pH 4.8, plus the cations to remove the glucose and cellobiase.

(iv) Finally, resuspend the slurry in 0.5 ml of the same buffer and determine its activity as described previously.

Table 8. Preparation of Cellobiase Adsorbed onto Con A-Sepharose.

Cellobiase added (units)	Cellobiase in washings (units)	Bound cellobiase (A) maximum (units)	Activity of complex (B) (units)	B/A (%)
18	0	18 ·	20	111

We have found that at these concentrations of glucose and cellobiose, desorption of enzyme activity does not occur (1,9).This inability to desorb has important implications because cellobiase will remain firmly bound to CAS in the presence of high substrate and product concentrations. In related investigations, we have also found that cellobiase activity does not desorb from the support between pH 3.5 and 7.4 (9).

Table 8 shows that no cellobiase activity is lost upon its adsorption to CAS. The reason for this may be because the adsorption mechanism does not involve the protein moiety of the enzyme (unlike adsorption onto ion-exchange resins) and thus the protein moiety containing the active site of the enzyme is less likely to be affected by the immobilisation procedure.

4. COVALENT COUPLING TO SEPHAROSE, CELLULOSE AND POLY-ACRYLAMIDE

The covalent linkage of an enzyme to a support material such as those given in *Table 2* requires that the support material be 'activated' so that it can then combine chemically with the enzyme. This method of immobilisation is exemplified by the covalent coupling of cellobiase to Sepharose and of invertase to cellulose and polyacrylamide.

4.1 Coupling of Cellobiase to Cyanogen Bromide-activated Sepharose

The covalent binding of an enzyme to cyanogen bromide-activated (CNBr) Sepharose is a widely used method for this kind of immobilisation. This material can be purchased commercially, and the enzyme can be coupled directly without the need for the activation step. A description of the preparation of CNBr-Sepharose will not be given here. For this, the reader is referred to reference 5. The activation procedure is based upon the reaction between cyanogen bromide and the hydroxyl groups of Sepharose. The resulting imidocarbonate groups can react with the free amino groups of the enzyme to be immobilised.

The enzyme cellobiase can be coupled to the support through the amino groups of its protein moiety or through its carbohydrate moiety since this enzyme is a glycoprotein (*Table 9*). However, glycoproteins can be difficult to immobilise by covalent coupling because the amino acid side chains containing free amino groups are made inaccessible to the activated support by the carbohydrate molecules (10). Also, immobilisation through the carbohydrate moiety offers the possibility that little effect upon enzyme activity will result. This is because the protein moiety, containing the catalytic active site, could be largely unaffected. This is generally found not to be the case if an enzyme is coupled to the support

Table 9. Covalent Immobilisation of Cellobiase on CNBr-activated Sepharose.

1. Take 5 g of CNBr-Sepharose and place on a sintered glass filter. Swell the gel by washing with 1 litre of 1 mM HCl and, finally, wash it thoroughly with 0.1 M NaHCO$_3$ buffer, pH 8.3, containing 0.5 M NaCl. 1 g of dry material swells and gives about 3.5 ml final gel volume.

2. Take 1.0 ml of cellobiase diluted 10 times with distilled water and treat the enzyme with sodium metaperiodate (0.02 M final concentration) for 1 h at room temperature (~23°C) in the dark.

3. Stop the reaction by the addition of 1.0 ml of 0.36 M ethylene glycol.

4. Dialyse the solution thoroughly against distilled water at 4°C for 7 h and then add an equal volume of 0.1 M sodium phosphate buffer pH 6.0 containing 0.2 mg/ml of sodium cyanoborohydride and 0.1 M ethylenediamine. Leave the mixture overnight at 4°C.

5. Remove excess reactants by dialysis against 10 mM sodium acetate buffer (pH 4.8) at 4°C[a].

6. The aminoalkylated carbohydrate side chains of cellobiase can now be coupled to CNBr-Sepharose. Mix 0.6 ml of the derivatised enzyme with 2.4 ml of swollen CNBr-Sepharose in 0.1 M NaCHO$_3$ buffer, pH 8.3, and mix overnight at 4°C.

7. Filter the enzyme-Sepharose complex, wash it thoroughly in 10 mM sodium acetate buffer, pH 4.8, and resuspend it in the same buffer. Enzyme activity should be monitored throughout the above procedure.

[a]The mechanism of the reaction is depicted in *Figure 1*.

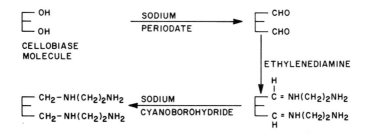

Figure 1. Aminoalkylation of the carbohydrate side chains of cellobiase.

Table 10. Preparation of Aminoalkylated Cellobiase Immobilised on CNBr-Sepharose.

Cellobiase added (units)	Cellobiase in washings (units)	Bound cellobiase (A) maximum (units)	Activity of complex (B) (units)	B/A (%)
4	0	4	3.2	81

directly through the protein moiety.

Note that aminoalkylation of an enzyme may result in a loss of activity. In the case of cellobiase, 33% of the original activity was lost (9). The preparation of aminoalkylated cellobiase covalently bound to CNBr-Sepharose is summarised in *Table 10*.

4.2 Coupling of Invertase onto Microcrystalline Cellulose

This method is based upon the activation of natural supports (e.g., cellulose) with

Table 11. Covalent Immobilisation of Invertase onto Cellulose.

1.	Activate 0.5 g of microcrystalline cellulose by shaking it (end-to-end rotation) with 10 ml of titanous chloride solution (15% w/v) for 3 h at room temperature.
2.	Wash the support exhaustively in 10 mM sodium acetate buffer, pH 4.7, and resuspend in 9.6 ml of the same buffer. The nature and pH of the buffer used will, of course, depend upon the properties of the enzyme to be immobilised.
3.	Add 0.4 ml of commercial invertase to the activated support (9.6 ml) and mix overnight at 4°C using rotation on rollers.
4.	Centrifuge the enzyme-support complex, remove the supernatant, wash the complex thoroughly with buffer, pH 4.7 (washings), and finally resuspend the complex in the same buffer to give a total volume of 10 ml.

Table 12. Preparation of Invertase Immobilised on Microcrystalline Cellulose.

Invertase added (units)	Invertase in washings (units)	Bound invertase (A) maximum (units)	Activity of complex (B) (units)	B/A (%)
1600	1000	600	275	46

the salts of transition metals (e.g., titanous chloride), which can be purchased commercially from chemical companies such as British Drug Houses and Aldrich Chemical Company. The immobilisation of biocatalysts by metal link/chelation processes is described in detail in Chapter 2. The procedure used is given in *Table 11*).

Note that the microcrystalline cellulose-enzyme complex should be evenly suspended prior to the removal of an aliquot for the determination of activity. Also, any enzyme that may have been adsorbed on cellulose can be removed by washing the complex in buffer containing 1.0 M NaCl. The preparation of invertase covalently bound to microcrystalline cellulose is summarised in *Table 12*.

4.3 Coupling of Invertase onto Polyacrylamide

Examples of polyacrylamide supports which are commercially available (Koch-Light Laboratories, Ltd.) are the Enzacryl type of matrices. These supports are co-polymers of acrylamide and various derivatives of acrylamide. Two examples are Enzacryl AA and AH, which contain the aromatic residue and acid hydrazide residue, respectively. After activation, they can covalently bind proteins, as depicted in *Figure 2*.

4.3.1 *Immobilisation of Commercial Invertase to Enzacryls AA and AH*

This procedure is given in *Table 13*. Enzacryls AA and AH readily bind invertase, but the bound enzyme on each support is inactive (11). They bind 64% and 46% of the initial enzyme activity, respectively, but all activity is lost on coupling. The reason for the total loss of activity is unknown. It is unlikely, however, that the total loss of activity seen when invertase is immobilised on these supports will be a general phenomenon occurring with all enzymes, so the methodology described above will be applicable.

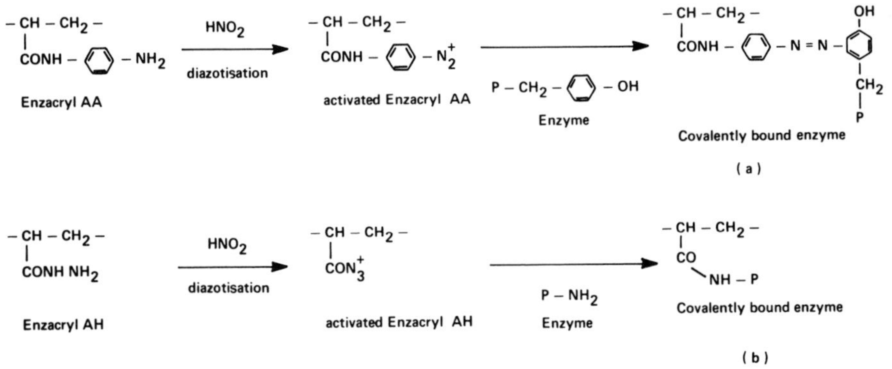

Figure 2. Activation of Enzacryls AA and AH followed by covalent binding of protein.

Table 13. Covalent Immobilisation of Invertase onto Polyacrylamide.

1.	Mix 100 g of Enzacryl AA or AH with 5 ml 2 N HCl at 4°C for 15 min (mixing performed by rotation on rollers).
2.	Add 4 m of ice-cold 2% (w/v) sodium nitrite solution, and continue the mixing for an additional 20 min.
3.	Wash the diazotised Enzacryls AA and AH thoroughly with 10 mM sodium phosphate buffer (pH 8.0) and 10 mM sodium carbonate (pH 9.2), respectively, and finally, resuspend them in 9.9 ml of the respective buffers.
4.	Add 0.1 ml of commercial invertase to 9.9 ml of the activated Enzacryl and mix for 48 h at 4°C by rotation.
5.	Centrifuge the Enzacryl-enzyme suspensions and remove the supernatants. Resuspend them in 10 ml of 10 mM sodium acetate buffer, pH 4.7, containing 1 M NaCl, and shake them vigorously for 5 min at room temperature.
6.	Finally, resuspend the Enzacryl-enzyme complexes in 10 ml of buffer, pH 4.7, and assay an aliquot for enzyme activity.

5. PROPERTIES OF ADSORBED AND COVALENTLY BOUND ENZYMES

From a practical point of view, the immobilisation of an enzyme onto a water-insoluble support allows the enzyme to be recovered after a reaction and re-used. It is absolutely essential to know that the immobilised enzyme remains firmly bound to the support under various conditions (e.g., pH, temperature, presence of substrate, and product) to which it may be subjected for maximum recovery of the enzyme. Also, once this secure binding has been established, a comparison of the properties of the free and immobilised enzyme can be made.

5.1 Maximum Loading of Enzyme to a Support

The maximum loading of an enzyme to a support material refers to the maximum amount of enzyme activity that can be immobilised on a certain amount of support. It is important to know this information because high loading of an enzyme onto a support material will minimise the space used in a reactor. The determination of maximum loading is exemplified by the immobilisation of cellobiase onto

Table 14. Maximum Loading of Cellobiase on CAS[a].

Test No.	Cellobiase added		Cellobiase in washings		Activity of immobilised enzymes		Activity $C/(A - B)$
	[units (A)]	(mg protein)	[units (B)]	(mg protein)	[units (C)][b]	(units/g CAS)	(%)
1	18	1	0	0.3	20	51	111
2	180	10	0	2.8	216	531	120
3	900	50	219	42.1	765	1880	112
4	1800	100	905	86.8	834	2050	93

[a]Reprinted from Lee and Woodward (1) with permission. Copyright, 1983, John Wiley and Sons Inc.
[b]Units refer to the amount contained in 10 ml of CAS slurry.

Con A-Sepharose. The data in *Table 14* show that the amount of cellobiase bound to the CAS slurry (see Section 3) can be increased several fold by increasing the amount of enzyme initially contacted with the support. Thus, the maximum loading of cellobiase activity upon the support is 83.4 units/ml, or 2050 units/g of dry CAS. It is also important to determine whether the specific activity (units/mg protein bound) is reduced as the loading of enzyme upon the support increases. This determination is made by measuring the amount of protein bound to the support as well as the amount of activity. A dramatic decrease in specific activity at high enzyme loading would mean that the enzyme is less efficient. This inefficiency may exist because proteins other than the enzyme in question are preferentially immobilised, or that crowding of protein on the support reduces the accessibility of the enzyme molecules to the support. From the data in *Table 14*, the specific activity of cellobiase on CAS is greater at a loading of 1880 units/g of CAS.

5.2 Kinetic Parameters of Immobilised Enzymes

The effect of pH, temperature, ionic strength, substrate and product concentration on the immobilised enzyme preparation are examples of kinetic parameters that should be determined and compared with data obtained for the free enzyme. These studies should be carried out for the free and immobilised enzyme in the same way. If the experiment is done in batch fashion (e.g., the enzyme is incubated with the substrate in a test tube), the rate of reaction observed with either the free or immobilised enzyme is measured under different conditions.

Predictions of how a particular property of an enzyme will be modified upon its immobilisation can sometimes be made, particularly when the enzyme is adsorbed onto a charged support. The pH activity and stability of an ionically adsorbed enzyme can be displaced either towards more alkaline or acid pHs if the sorbent is negatively charged or positively charged, respectively. Such displacements are attributed to the microenvironment in the immediate vicinity of the charged support and enzyme. Such effects can be seen when yeast invertase is immobilised on DEAE- and CM-Sephadex (*Figure 3*).

The determination of the K_m value of an enzyme is initially important, and

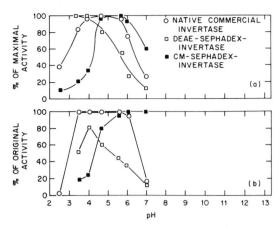

Figure 3. The effect of pH on the activity (a) and stability (b) of DEAE-Sephadex and CM-Sephadex-invertase complexes.

immobilisation of an enzyme can result in the increase or decrease of this parameter. A decrease in the K_m value of an enzyme upon immobilisation (the resulting value is referred to as the apparent K_m) can have practical advantages because the rate of reaction will be faster at lower substrate concentrations. Conversely, an increase in K_m upon immobilisation means that a higher substrate concentration is required to achieve the same rate of reaction observed with the free enzyme. Again, changes in this parameter can be attributed to microenvironmental effects in the vicinity of the immobilised enzyme particles. For example, surrounding the particle is an unstirred layer (Nernst layer) of solvent in which the concentration of substrate will be lower than that in the bulk solution. Consequently, a higher substrate concentration will be required in the bulk solution to saturate the immobilised enzyme with substrate. This effect can be demonstrated by reducing the size of the particle to which the enzyme is immobilised or by increasing the stirring speed of the particles, which reduces the apparent K_m value. The ionic nature of the support material can also influence K_m values, particulary when the substrate is itself charged. The apparent K_m of an immobilised enzyme will be reduced if the charges on the support and substrate are opposite. If the charges are alike, the apparent K_m will be increased. Other cases of increases or decreases in K_m are more difficult to explain. The apparent K_m values of invertase and cellobiase immobilised on various supports are given in *Table 15*.

It is hoped that the immobilisation of an enzyme on a support will result in an increase of the thermal stability of an enzyme which, of course, will prolong the activity of the enzyme. It is difficult to predict, however, whether a particular method of immobilisation will result in such an increase. The effect of temperature on the activity of the enzyme is determined by measuring the initial rate of the reaction at different temperatures. An increase in the temperature optimum of the immobilised enzyme does not necessarily mean that its temperature stability will be increased. Although the temperature optimum may increase by $10-20°C$, the immobilised enzyme may be very unstable at the higher

Table 15. Apparent K_m Values[a] of Immobilised Invertase and Cellobiase Preparations.

Type of support	Invertase	Cellobiase
None	28.0	2.7
Con A-Sepharose	28.5	1.7
Microcrystalline cellulose	28.5	1.1
CNBr-Sepharose	–	1.2
CM-Sephadex	100.0	–

[a] Values are expressed in mmol/l (1)

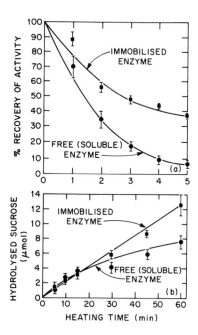

Figure 4. Thermal stability (65°C) of soluble and Con A-agarose-immobilised invertase in (**a**) the absence and (**b**) the presence of 0.12 M sucrose.

temperature and lose its activity rapidly even though the initial rate of reaction is increased. Temperature stability can be measured in two ways:

(i) by pre-heating the enzyme for various times, cooling, followed by activity determination; and

(ii) by incubating the enzyme and substrate together at a particular temperature and monitoring the conversion of substrate to product.

The data shown in *Figure 4* give an example of both. In these data, invertase is seen to have increased stability when it is adsorbed on CAS or covalently bonded to microcrystalline cellulose. *Figure 4b* shows that the immobilised enzyme preparations are actually more stable during operation than the free enzyme. Such data are more useful than those in *Figure 4a*.

6. ACKNOWLEDGEMENTS

The author acknowledges the secretarial assistance of Ms. D.J. Weaver. This work was supported in part by the Office of Basic Energy Sciences, Department of Energy under Contract No. DE-AC05-840R21400 with Martin Marietta Energy Systems, Inc.

7. REFERENCES

1. Lee,J.M. and Woodward,J. (1983) *Biotechnol. Bioeng.,* **25**, 2441.
2. Zaborsky,O. (1973) *Immobilised Enzymes*, published by CRC Press, Cleveland, Ohio.
3. Barker,S.A. and Kay,I. (1975) in *Handbook of Enzyme Biotechnology*, Wiseman,A. (ed.), Ellis Horwood, Chichester, p. 89.
4. Goldstein,G. and Manecke,G. (1976) in *Immobilised Enzyme Principles*, Wingard,L.B.,Jr., Katchalski-Katzir,E. and Goldstein,L. (eds.), Academic Press, New York, p. 23.
5. Mosbach,K., ed. (1976) *Methods in Enzymology,* Vol. **46**, published by Acdemic Press, New York.
6. Woodward,J. and Wiseman,A. (1978) *Biochim. Biophys. Acta,* **527**, 8.
7. Woodward,J. and Wiseman,A. (1982) in *Developments in Food Carbohydrate-3*, Lee,C.K. and Lindley,M.G. (eds.), Applied Science Publishers, Ltd., Barking, p. 1.
8. Woodward,J. and Wiseman,A. (1982) *Enzyme Microb. Tech.,* **4**, 73.
9. Woodward,J. and Wohlpart,D.L. (1982) *J. Chem. Technol. Biotechnol.,* **32**, 547.
10. Hsiao,H.-Y. and Royer,G.P. (1979) *Arch. Biochem. Biophys.,* **198**, 379.
11. Woodward,J. (1977) Ph.D. Thesis, University of Surrey.

CHAPTER 2

Immobilisation of Biocatalysts by Metal-Link/Chelation Processes

JOHN F. KENNEDY and JOAQUIM M. S. CABRAL

1. INTRODUCTION

The overall principle of attachment of a biologically active molecule to an insoluble matrix is simple and simulates the natural mode of action and environment of enzymes, antibodies, antigens, etc., which are carried on the surface or in the interior of cells or which are embedded in biological membranes and tissues.

A number of methods of immobilisation of biological molecules exists and no one method is perfect for all molecules or purposes. When attaching a biologically active molecule to an insoluble support, it is important to avoid a mode of attachment that reacts with or disturbs the active site(s) of the molecule, as otherwise a loss of activity will result on binding. It is also important to avoid overloading the matrix when binding molecules, since overloading leads to overcrowding and hence reduced activity, by reason of steric hindrance of approach of the substrate molecules to the active sites of the bound molecules. However it does not follow that limited loading of the enzyme molecules on the matrix surface will be successful, since hydrogen bonding and hydrophobic forces may occur between immobilised enzyme and 'free' molecules thus causing the latter to block up the spaces between the immobilised molecules. Attention to the way in which the macromolecule can be attached to the insoluble matrix and the choice of matrix is also a matter of importance. Methods of insolubilisation of enzymes, cells, antigens, antibodies, nucleic acids, antibiotics and affinants, and descriptions of the insolubilised molecules have been reviewed (1,2).

Covalent-type linkages between molecule and matrix are generally best but many of the recommended immobilisation procedures require pre-derivatisation of the matrix, long coupling reactions and specialised conditions. We were searching for a method in which matrix derivatisation after preparation could be avoided, and where virtually instantaneous coupling could be achieved under simple conditions. It occurred to us that the chelation properties of transition metals could be employed to this end. While there is a wide choice of transition metals, the properties of titanium and zirconium seemed particularly attractive on account of the non-toxicity of their oxides.

2. IMMOBILISATION OF ENZYMES BY METAL-LINK/CHELATION PROCESSES

2.1 Enzyme Immobilisation Techniques on Transition Metal-activated Supports

2.1.1 *Original Metal-link Method*

The metal-link method for immobilisation of enzymes was initially developed by Novais (3) at the University of Birmingham, UK, while carrying out some experiments on an immobilisation enzyme technique on cellulose by diazotisation. Titanium(III) chloride was used in a reducing step of a nitroaryl derivative of cellulose [2-hydroxy-3-(*p*-nitrophenoxy)propylether cellulose] to yield the corresponding aminoaryl derivative which promotes the diazo coupling of the enzyme. The inclusion of this transition metal chloride led to immobilised enzyme preparation whose activities were higher than in its absence. The idea was then to activate the support (cellulose) only with the transition metal salt. The results obtained showed clearly their dependence on the concentration of the transition metal chloride used on the activation of cellulose. The technique is quite simple and comprises only two steps, as shown in *Table 1*.

2.1.2 *Chemistry of the Enzyme Coupling*

The way in which transition metal compounds chelate biopolymers, etc., is illustrated from the viewpoint of a simple system of titanium(IV) chloride and cellulose.

A proportion of the titanium ions in the titanium chloride solution is octahedrally coordinated with molecules or ionic species that are essentially the ligands of the complex ion (see *Figure 1* for examples of complex ions). Hydroxy groups are effective ligands for the transition metal ions, and therefore it is to be expected that transition metal ions may complex with polysaccharides in which the hydroxy groups act as new ligands replacing others. Moreover, it is well known that glycols are very effective ligands that combine with transition metal ions. Certain polysaccharides such as cellulose contain vicinal diol groups not involved in the glycosidic linkages between residues and therefore are amenable to chelation by transition metal ions, the chelate being formed by replacement of two of the titanium ion ligands by polysaccharide hydroxy groups. Thus, since

Table 1. Activation of Support and Coupling of Enzyme.

1.	Weigh 10 g of microcrystalline cellulose onto a dish.
2.	Add 50 ml of a 15% w/v titanium(IV) chloride solution in 15% w/v HCl and mix well.
3.	Place the mixture in a desiccator containing sodium hydroxide and evacuate.
4.	Dry the evacuated mixture in an oven at 45°C until complete dryness (⊙ 24 h).
5.	Wash the solid three times with buffer (at a pH which is usually optimum for enzyme activity) until off-white in color.
6.	Add to the washed solid an enzyme solution, containing 1 g of protein, and stir at 4°C for 18 h.
7.	Centrifuge the suspension and wash the immobilised enzyme with two cycles of buffer, and sodium chloride (1 M) in the same buffer, followed by two washings with the buffer alone.
8.	Resuspend the immobilised enzyme in the same buffer and store at 4°C.

$[Ti(H_2O)_6]^{4+}$

L = neutral ligand

$[Ti(H_2O)_5Cl]^{3+}$ $[Ti(H_2O)Cl_5]^{-}$ $[TiCl_6]^{2-}$

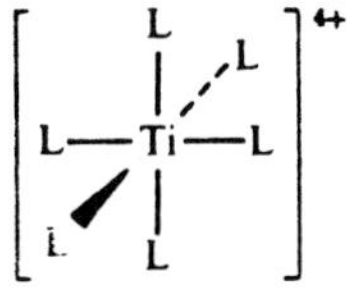

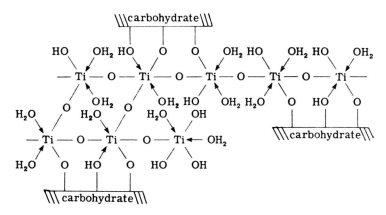

Figure 1. Example of complex ions.

Figure 2. Titanium-treated cellulose chelated and/or complexed along the length of its chains with various aquo, chloroaquo, and chloro complex titanium species prevalent in solution.

cellulose is a polymer of β-1,4-linked D-glucopyranose units, the 2- and 3-hydroxy groups would be involved in chelate formation. The 6-hydroxy group of the D-glucopyranose residue can only be expected to participate in complexing of the cellulose chain by titanium, since sterically this hydroxy group is unable to become sufficiently close to any of the other hydroxy groups of the cellulose chain to be involved in chelate formation. Thus, titanium-treated cellulose material may be envisaged as being chelated and/or complexed along the length of its chain with various aquo, chloroaquo, and chloro complex titanium species prevalent in solution (*Figure 2*).

It should be noted that, for steric reasons, it is impossible for all the water or chloride ligands of the titanium ion to become replaced by other hydroxy groups of the polysaccharide chain. Furthermore, these remaining ligands will not be extensively replaced by hydroxy groups of adjacent cellulose chains on account of the insolubility and lack of mobility of the cellulose molecules. Chelation may be achieved for any molecule containing groups appropriate to replace the ligands of the titanium bound to cellulose. The incoming molecule should be in aqueous solution and a near-neutral pH is expected to be adequate for the coupling. Thus, for proteinaceous molecules such as enzymes, antibodies and antigens, the molecular types most frequently immobilised, groups that can act as ligands will

Table 2. The Use of Transition Salts for the Coupling of Glucoamylase to Inert Surfaces.

Support material	Metal salt activator	Salt concentration (% w/v)	Activity[a] (U/g)	Bound protein (mg/g)	Specific activity retention (%)
Microcrystalline cellulose	TiCl$_4$	15.0	3300	95.5	52.0
		7.5	3020	93.5	49.0
		3.0	2800	92.0	46.0
		1.5	2600	85.0	46.5
		0.3	775	26.6	43.5
		0.15	665	13.8	72.5
	TiCl$_3$	12.5	4570	147	46.5
		6.25	3380	106	44.0
		2.5	2600	88.5	45.5
		1.25	2350	69.5	51.0
	SnCl$_4$	10.0	2100	58.8	56.0
		1.0	1080	20.4	83.0
	SnCl$_2$	10.0	1325	44.9	45.0
	ZrCl$_4$	10.0	2970	88.8	50.0
		0.1	1192	43.5	41.0
	VCl$_3$	10.0	900	22.3	60.5
	FeCl$_3$	10.0	1430	35.0	61.5
	FeCl$_2$	10.0	1470	39.4	56.0
		0.1	400	8.1	74.0
Glass (ground borosilicate)	TiCl$_4$	15.0	2860	79.0	52.6
	TiCl$_3$	12.5	2300	68.5	50.0
	SnCl$_4$	10.0	1090	29.6	47.2
Nylon (66 fibre)	TiCl$_4$	15.0	2780	n.d.	n.d.
		7.5	1525	n.d.	n.d.

[a]One glucoamylase unit liberates reducing sugars equivalent to 1 μmol of glucose at 45°C in 1 min from 1% w/v starch solution in 0.02 M acetate buffer, pH 4.5.
n.d. = not determined.

be the free carboxy groups from the C terminus of acidic amino acids, the phenolic hydroxy groups of tyrosyl residues, the alcoholic hydroxy groups of seryl and threonyl residues, free sulphydryl groups from any cysteinyl residues and amino groups from the N terminus of ϵ-amino groups of lysyl residues.

The production of such a bond between the titanium-treated polymer and the enzyme, to produce an enzymically active derivative, would be expected to depend on:

(a) the availability in the enzyme molecule of groups which can act as ligands;
(b) steric factors which permit such groups to come in contact with the titanium atom;
(c) the non-involvement of such groups in the region of the active site(s) of the enzyme; and
(d) the close proximity of enzyme molecules already bound in the polymer matrix.

Table 3. Comparison of Specific Activity Retention of Enzymes Chelated to Titanium(IV)-activated Supports.

Enzyme	Support material	Specific activity retention (%)	Reference
α-Amylase	Cellulose	54.0	4
Glucoamylase	Cellulose	46 – 72.5	4,5
	Glass (ground borosilicate)	56.2	5
	Porous glass (pore size 100 Å)	20.0	6
	Spheron	60 – 82	7
D-Glucose oxidase	Cellulose	61.5	4
Invertase	Cellulose	3.6	4
Nuclease P1	Porous glass (pore size 208 Å)	52.1	8
(3′AMP-dephosphorylating	Pumice stone	66.2	8
activity)	Silica gel (pore size 85.5 Å)	46.1	8
	Silica gel (pore size 190 Å)	63.2	8
(nucleolytic activity)	Porous glass (pore size 208 Å)	36.0	8
	Pumice stone	52.1	8
	Silica gel (pore size 87.5 Å)	13.5	8
	Silica gel (pore size 190 Å)	42.1	8
Papain	Glass fibre		
(esterolytic activty)		76.6	9
(proteolytic activity)		37.9	9
Trypsin	Spheron P 100 000		
(esterolytic activity)		19.0	7
(proteolytic activity)		0.46	7
Urease	Cellulose	70.0	5
	Nylon 66	81.0	5

The possibility of ionic binding is discounted because the enzymes can be bound to the titanium complexes of the polymers in ionic media.

2.1.3 *Properties of Metal-linked Enzymes*

Table 2 shows some typical results (3) for a selection of supports obtained with various transition metal salts used to couple glucoamylase. The applicability of this method to a number of enzymes is illustrated in *Table 3*. From these tables it can be seen that the specific activity retention of the transiton metal-chelated enzymes is usually high (50 – 80%). This immobilisation technique is also reported not to modify the Michaelis constant of the enzyme after immobilisation (*Table 4*).

The pH for optimum activity of the chelated enzyme remains, in several cases, unaltered, as can be seen in *Table 5*; however, shifts in pH of 0.5 – 1.0 pH units were observed either in the acid or alkaline direction.

A survey of the published work on the application of this method to coupling enzymes to various supports reveals that widely varying stabilities may be obtained. Moreover, stabilities have often been inadequately stated, in the sense that only storage stabilities were examined. The operational stability is one of the most

Table 4. Comparison of Michaelis Constant (K_m) of Enzymes Chelated to Titanium(IV)-activated Supports.

Enzyme	Support material	K_m Soluble enzyme	Immobilised enzyme	Reference
D-Glucose-isomerase	Cellulose	25.6% w/v	25.3% w/v	10
Invertase	Hornblend	50 mM	50 mM	11

Table 5. Comparison of Optimum pH Activity of Enzymes Chelated to Titanium(IV)-activated Supports.

Enzyme	Support material	Optimum pH Soluble enzyme	Immobilised enzyme	Reference
Glucoamylase	Cellulose	4.5	4.5	5
D-Glucose-isomerase	Cellulose	7.0	6.0	10
Invertase	Hornblend	5.0	4.0	11
Papain	Glass fibre	6.0	6.5	9
Nuclease P1	Pumice stone	5.3	5.3	8

Table 6. Operational Stability of Enzymes Immobilised on Titanium(IV)-activated Supports.

Enzyme	Support	Operational temperature (°C)	Half-life	Reference
Glucoamylase	Cellulose	50	10 h	5
	Porous glass	45	1 h	6
Invertase	Hornblend	25	54 days	11
	Porous glass	18	8 days	11

important tests to assess the potential industrial use of an immobilised enzyme. The operational stabilities of metal-chelated enzymes are shown in *Table 6*.

The best half-lives were only obtained at relatively low temperatures for an enzyme, invertase, which is very stable at the temperature tested. Furthermore, glucoamylase, a stable enzyme, when immobilised by chelation shows a very unstable behavior.

2.1.4 *Problems of the Metal-link Method*

Although the original metal-link method for enzyme immobilisation is a quite simple technique, several authors (12,13) report non-reproducible results usually with inorganic supports. The interaction of the transition metal salt, namely titanium(IV) chloride, with the silanol groups of the inorganic supports is not very strong (14,15) and chelation only occurs to a small extent, depending on the hydrophilicity of the support, that is, on the free silanol groups available on the surface of the inorganic supports.

Using chemical evidence, Kennedy (9) and Cabral (15,16) have proposed that, with inorganic supports, a layer of hydrous metal oxide on the support surface is responsible for the immobilisation of the enzymes rather than a chemical inter-action between the transition metal and the support. This layer is thought to be obtained by drying the support in the presence of excess salt solution. During heating, the evaporation of water and hydrogen chloride occurs and the for-mation of hydrous metal oxide and/or oxychloride is induced with concomitant precipitation on the support. A stable layer is, however, obtained only if the mix-ture of support and transition metal solution is completely dried since, in the sub-sequent washing with water, the fairly well bound hydrous metal oxide can become removed (16).

The complete dryness of the mixture of the transition metal salt solution with the support is the most important detail in this technique. The dryness (activa-tion) temperature is also critical in order to obtain highly active immobilised enzyme preparations. With an increasing temperature, although a more stable and easily formed metal oxide layer is obtained, the chelation degree of the enzyme molecules decreases, as does the hydrophilic character of the hydrous metal oxide, and so there are less active sites on the metal oxide surface to chelate the enzyme molecules.

Other important parameters are the ratios of transition metal salt solution to support weight and of activation time to support weight (or volume). The best results are obtained when the activation is carried out by drying the support, mainly with inorganic matrices, in the presence of excess salt solution, as in this case a maximum hydrous metal oxide layer is obtained. With porous supports however, this ratio must be optimised in order to give the maximum active immobilised enzyme preparation, as in the presence of excess transition metal salt the pores of the support may be partially obstructed leading to a lower surface area on which to couple the enzyme molecules (16,17).

The activation time of the support with the transition metal salt solution is usually reported to be 'overnight' or 24 h; however, this parameter depends on the weight or volume of the support to be activated, as the height of the slurry support and activation solution may change. However, if the height of this mix-ture is kept between 3 and 5 mm, using appropriate containers, the activation

Table 7. Influence of the Drying Conditions on the Operational Stability and Composition of Glucoamylase Immobilised on a Hydrous Titanium(IV) Oxide Derivative of Controlled Pore Glass.

Time (hours)	Activity (U/g matrix)		Bound protein (mg/g matrix)		Specific activity (U/mg protein)		Bound titanium (mg $TiCl_4$/g matrix)		Protein/titanium ratio (mg/mg $TiCl_4$)	
	A	B	A	B	A	B	A	B	A	B
0	71.5	239	3.45	18.5	20.8	12.9	23.2	98.0	0.153	0.189
1	–	105	–	11.5	–	9.13	–	95.6	–	0.120
24	40.0	62.7	3.41	11.1	11.7	5.64	22.2	95.0	0.154	0.117
48	33.6	45.7	3.37	10.8	9.97	4.24	20.5	94.7	0.164	0.114
72	28.5	38.5	3.14	11.1	9.09	3.47	21.7	94.5	0.145	0.117
96	–	26.4	–	10.7	–	2.46	–	94.1	–	0.114
120	25.6	17.6	3.11	9.56	8.25	1.84	20.6	93.2	0.151	0.103

A = Drying at atmospheric pressure.
B = Drying under vacuum.

time can be kept a constant (24 h) in the procedure (16).

Vacuum can also be used to assure a higher titanium level on the activated support (15), as can be seen in *Table 7*. When the activation is performed under vacuum, a higher titanium (98.0 mg $TiCl_4$/g matrix) level is obtained than at atmospheric pressure (23.2 mg $TiCl_4$/g matrix). This is probably due to an incomplete dryness of the mixture support plus $TiCl_4$ solution which, in contact with water, is easily removed as a hydrous titanium(IV) oxide. Even when the drying is performed under vacuum, the titanium(IV) layer obtained on the surface of the support is only 6.5% of the initial amount of $TiCl_4$ which is put into contact with the support (1500 mg $TiCl_4$/g matrix).

During the first hours of operation a loss of titanium is observed from the preparation activated under vacuum. This loss is partially responsible for the decrease of enzyme activity, although there is a loss of 'free' enzyme, not bound to titanium, as can be seen by the decrease in protein/titanium ratio during the first hour of operation (*Table 7*).

With the support at atmospheric pressure a slight decrease in the protein level with a corresponding decrease in the titanium level on the support is observed. This leads to a protein/titanium ratio approximately constant for the whole operational time.

Table 8. Activation of Inorganic Supports with Transition Metal Salts (Lehigh).

1.	Weigh 10 g of inorganic support (porous alumina, stainless steel, Ni-NiO) and put in a reaction flask.
2.	Partially immerse the reaction flask in an ice bath and add 62.5 ml of chilled water.
3.	Add slowly and with vigorous stirring 6.3 ml of pure liquid titanium(IV) chloride to the mixture of support and chilled water.
4.	Remove the flask from the ice bath and heat at 80°C and hold it at this temperature for 1 h.
5.	Remove the resulting hydrous titanium(IV) oxide-coated support and wash thoroughly with tap water until a clean rinse is obtained.
6.	Dry the washed and coated support overnight at 110°C.
7.	Add an enzyme solution containing 1 g of protein in deionised water to the coated support and contact at $0-5$°C for 18 h.
8.	Remove the supernatant solution and wash the immobilised enzyme four to six times with deionised water before assaying for enzyme activity.

Table 9. Activation of Inorganic Supports with Transition Metal Salts (Galway).

1.	Weigh 2 g of inorganic support (hornblend, porous glass) and put in a 25 ml screw-cap bottle.
2.	Add 10 ml of a 10% w/v titanium(IV) chloride solution and heat the mixture at 30°C for 1 h.
3.	Remove the titanium(IV) chloride residual solution.
4.	Wash the activated support thoroughly with distilled water.
5.	Dry the washed support overnight at 100°C.
6.	Add an enzyme solution containing $0.5-4.0$ g of protein in standard buffer (for example, for glucoamylase or invertase, 0.2 M sodium acetate, pH 4.5) to the activated support and bind the enzyme at 4°C for 16 h.
7.	Remove the unbound enzyme solution.
8.	Wash the immobilised enzyme with 1.0 M NaCl in standard buffer followed by standard buffer alone.

Although the protein level is approximately constant, after the first 24 h a decrease in the specific enzyme activity is always observed for both procedures. These results, as well as those obtained by Flynn *et al.* (18) and Cabral *et al.* (6), show the influence of the surface of the titanium(IV)-activated supports on the parameters which control the deactivation of glucoamylase, probably due to a type of inhibition of the metal activator.

2.2 Developments of the Metal-link Method

The original metal-link method was applied mainly to polysaccharide supports, namely cellulose, which conferred the highest activty to immobilised enzyme preparations. The extension of this method to other supports (inorganic and proteinaceous) causes some problems regarding their use with the original technique. Thus, several procedures have been developed to achieve better results. Some of these techniques are described in the following section.

2.2.1 *Alternative Activation Techniques of Inorganic Matrices*

The procedures given in *Tables 8, 9* and *10* were developed at Lehigh (USA), Galway (Ireland) and Birmingham (UK) Universities, respectively.

Table 10. Activation of Inorganic Supports with Transition Metal Salts (Birmingham).

Heating procedure
1. Weigh 10 g of lead glass and put into a flask.
2. Add 40 ml of a 4% EDTA (disodium salt) solution at pH 10 for 48 h (this treatment will remove part of the lead-etched glass).
3. Wash the etched glass thoroughly with distilled water.
4. Add 40 ml of a 1:1 solution of $H_2SO_4:H_2O$ and leave to soak at room temperature for 5 days.
5. Wash thoroughly with distilled water.
6. Add 40 ml of a 15% w/v titanium(IV) chloride solution in 15% w/v hydrochloric acid and reflux the mixture for 3 – 5 h.
7. Remove the titanium(IV) chloride expent solution and wash thoroughly with distilled water.
8. Add an enzyme solution containing 1 g of protein in standard buffer (i.e., at pH optimum for activity) to the activated support and bind the enzyme at 4°C for 18 h.
9. Wash the immobilised enzyme with standard buffer (30 min stirring), 1.0 M NaCl in standard buffer (30 min stirring), 6 M urea solution (60 min stirring) and standard buffer (30 min stirring).
    Note: An optional step can be included to dry the activated support at 600°C for 8 h, before the enzyme coupling.

Neutralising procedure
1. Weigh 10 g of magnetic iron oxide [iron (II) di iron (III), Fe_3O_4].
2. Add 5 ml of pure (liquid) titanium(IV) chloride to the iron oxide and immediately add 100 ml of distilled water.
3. Neutralise the solution to pH 5.8 with 1.0 M ammonium hydroxide solution with constant stirring.
4. Remove the liquors (supernatant) by centrifugation.
5. Wash the activated support with standard buffer and add an enzyme solution (an undiluted solution of α-amylase culture filtrate from a *Bacillus* sp.).
6. After 1 h at 4°C, with frequent shaking, remove the unbound enzyme solution and wash the immobilised enzyme five times with the standard buffer.

At the University of Birmingham, two alternative techniques (*Table 10*) were developed achieving precipitation of hydrous metal oxide by heating (12) and by neutralisation (19,20).

The heating process was used to key a metal oxide layer on non-porous solid lead glass beads. In this process, the lead of glass was removed so that the structure would be opened up to facilitate deposition of the metal oxide layer within the bead surface. Leaching with 1:1 sulphuric acid solution followed by refluxing under titanium(IV) chloride solution was carried out to achieve the objective of the technique.

2.2.2 Activation of Proteinaceous Matrices

In most cases of enzyme immobilisation *via* transition metal chelation using titanium(IV) chloride, pure or in an acidic solution, the materials employed as supports have been inorganic aerogel-type (glass, silicon, etc.) or organic cross-linked polysaccharides aerogel-zerogel hybrids (e.g., cellulose derivatives). In such cases, activation with titanium(IV) chloride solutions, at low pH values, can be carried out easily without disruption of the support matrix.

In the case of more sensitive supports, namely those of proteinaceous nature (e.g., gelatin), the activation conditions should be as mild as possible so that the maximum possible level of active surface-bound enzyme is obtained under conditions of minimum deterioration of the support.

Table 11. Activation of Proteinaceous Matrices.

Titanium(IV) chloride-acrylamide complex

1. Add, with stirring, 400 ml of a solution of acrylamide (0.1 mol) in dry dichloromethane to 400 ml of titanium(IV) chloride (0.1 mol) in dry dichloromethane.
2. Stir for 10 min. On continuous stirring a solidified yellow complex is obtained from the viscous solution.
3. Collect this solid by filtration, grind, wash with dichloromethane and dry at room temperature.
4. Add 20 ml of an aqueous solution of titanium(IV) chloride-acrylamide complex (100 mg/ml) at pH 3.0 to a 2.0 ml bed volume of glutaraldehyde cross-linked gelatin particles.
5. Stir the mixture for 30 min at 18°C.
6. Remove the supernatant and wash the titanium(IV)-acrylamide activated gelatin particles with 0.1 M acetate buffer, pH 4.5.
7. Use immediately for coupling to the enzyme.

Titanium(IV) chloride-urea complex

This complex is prepared in a similar manner to the titanium(IV) chloride-acrylamide adduct.

1. Add, with vigorous stirring, 400 ml of a suspension of recrystallised urea (0.1 mol) in dry dichloromethane to 200 ml of a solution of titanium(IV) chloride (0.1 mol) in dry dichloromethane.
2. After 80 h collect a yellow solid by filtration. Grind, wash with dry dichloromethane and dry at room temperature.
3. Add 5.0 ml of an aqueous solution of titanium(IV) chloride-urea complex (40 mg/ml) at pH 3.0 to a 2.0 ml bed volume of glutaraldehyde cross-linked gelatin particles.
4. Stir the mixture for 60 min at 18°C.
5. Remove the supernatant and wash the titanium(IV)-urea activated gelatin particles with 0.1 M acetate buffer, pH 4.5.
6. Use immediately for coupling to the enzyme.

Table 11. (continued)

Titanium(IV) chloride-citric acid complex

1. Add 0.1 ml of 15% w/v titanium(IV) chloride in 15% w/v hydrochloric acid to 40 ml of a 0.05 M solution of citric acid. Adjust the pH to 4.5 with 5.0 M ammonia.
2. Add 50 ml of titanium(IV) chloride in citric acid solution, prepared as described above, to a 2.0 ml bed volume of glutaraldehyde cross-linked gelatin particles.
3. Stir the mixture for 60 min at 18°C.
4. Remove the supernatant and wash the titanium(IV)-citric acid activated gelatin particles with 0.1 M acetate buffer, pH 4.5.
5. Use immediately for coupling to the enzyme.

Titanium(IV) chloride-lactose complex

1. Add 2.0 ml of 15% w/v titanium(IV) chloride in 15% w/v hydrochloric acid to 18.0 ml of 0.5 M lactose. Adjust the pH to 3.0 with 50 M ammonia.
2. Add 5.0 ml of titanium(IV) chloride in lactose solution, prepared as described above, to a 2.0 ml bed volume of glutaraldehyde cross-linked gelatin particles.
3. Stir the mixture for 60 min at 18°C.
4. Remove the supernatant and wash the titanium(IV)-lactose activated gelatin particles with 0.1 M acetate buffer, pH 4.5.
5. Use immediately for coupling to the enzyme.

Activation of glutaraldehyde-gelatin matrix with uncomplexed titanium(IV) chloride

1. Add 1.0 ml of pure titanium(IV) chloride to 5.0 g of moist water-washed or 2.0 g of freeze-dried glutaraldehyde cross-linked gelatin particles.
2. Stir the mixtures occasionally for 0.5 – 4.0 min.
3. Remove the excess titanium(IV) chloride and wash the activated beads with 0.1 M acetate buffer pH 4.5 until the pH of the liquids is 4.5.
4. Use immediately for coupling to the enzyme.

Table 12. Influence of the Titanium(IV) Chloride-complex Immobilisation Method on the Activity of Surface-bound Gelatin Glucoamylase (Activation Time: 1 h)

Activating adduct	Activity[a] (U/g)
$TiCl_4$-acrylamide	1.03
$TiCl_4$-urea	0.36
$TiCl_4$-citrate	0.41
$TiCl_4$-lactose	0.48

[a]One glucoamylase unit librates 1 μmol of D-glucose per min from a 1% w/v soluble starch solution at 55°C and pH 4.5.

The major obstacle, as far as protein denaturation is concerned, encountered with the titanium(IV) chloride activation of a glutaraldehyde cross-linked gelatin matrix, is the very low pH values of the commercial solutions of titanium(IV) chloride in hydrochloric acid which are usually employed as the source of titanium(IV). Several mild techniques in which the titanium(IV) remains in a soluble and reactive form at high pH values and therefore gives rise to less disruption of proteins have been developed (21,22). These procedures are given in *Table 11*. The influence of the different procedures on the activity of glucoamylase immobilised on the surface of glutaraldehyde cross-linked gelatin particles is shown in *Table 12*.

Titanium(IV) chloride is used in an uncomplexed and, hence, more reactive form. When either in solution in hydrochloric acid or pure, for the activation of blank cross-linked gelatin particles high levels of activity of gelatin glucoamylase preparations are obtained (*Table 10*), mainly when freeze-dried gelatin particles are employed.

2.2.3 Activation of Ion-exchange Matrices

Another useful support, in bead form, which presents several hydrodynamic advantages for use in enzyme reactors is the cross-linked polystyrene ion-exchangers. The polystyrene sulphonate cation exchangers can be equilibrated with several transition metals (Ti^{4+}, Zr^{4+}, V^{5+}, etc.) and used to immobilise enzymes *via* chelation. A typical procedure is described in *Table 13*.

2.2.4 Activation of Nylon Tubes

An alternative procedure to the original metal-link method used for activating nylon tubes with transition metal salts ($TiCl_4$, $SnCl_4$, $MoCl_5$, and VCl_3) is shown in *Table 14*. In this technique, nylon tubes are superficially hydrolysed to provide carboxyl and ammonium groups on their surfaces, which can act as ligands of the transition metal.

2.3 Cross-linking of Transition Metal-chelated Enzymes

Owing to the low operational stabilities of enzymes immobilised on hydrous titanium(IV) oxide derivatives of inorganic matrices, a cross-linking step in the

Table 13. Activation of Ion-exchange Matrices.

1.	Measure 1 ml of polystyrene sulphonate cation exchangers (H^+ form) and suspend in 2 ml of a 20% w/v aqueous solution of titanium(IV) chloride. Shake for 60 min at 20°C.
2.	Remove the supernatant and wash the solid with 0.1 M acetate buffer, pH 5.0.
3.	Suspend the titanium(IV) equilibrated resin in 1 ml of nuclease P1 (2 mg/ml, in 0.1 M $ZnCl_2$, 0.1 M acetate buffer, pH 5.0). Shake the suspension for 60 min at 20°C.
4.	Remove the unbound protein solution.
5.	Wash the immobilised enzyme five times with 1 M NaCl in 0.1 M acetate buffer, pH 5.0, and then three times with 0.1 M acetate buffer, pH 5.0, on a glass filter.

Table 14. Activation of Nylon Tubes.

1.	Pump 50 ml of 10% w/v titanium(IV) tetrachloride in pentane through air-dried (45°C) nylon tubing (2.6 m x 1.9 m), which has been superficially hydrolysed, at a flow-rate of 1.7 ml/min (activation time: 30 min).
2.	Dry the tubing at 45°C for 2 h.
3.	Rinse the titanium(IV)-activated tube with 250 ml of 0.05 M Tris-HCl buffer, pH 8.0.
4.	Dissolve 20 mg of enzyme (glucose dehydrogenase, 260 U/mg protein) in 5 ml of the same buffer, containing 3.3 mg NAD.
5.	Circulate the enzyme solution through the tubing at 4°C for 24 h at a flow-rate of 1 ml/min.
6.	Wash the enzyme tubing consecutively with 100 ml of coupling buffer, 100 ml 1 M NaCl in 0.12 M phosphate buffer, pH 7.4, and 100 ml 0.5 M sodium citrate, pH 7.4.
7.	Store the empty immobilised enzyme tubing at low temperature when not in use.

immobilisation procedure can be introduced in order to decrease the protein loss into solution during operation.

2.3.1 *Glutaraldehyde Cross-linking*

A cross-linking agent frequently used is glutaraldehyde. The cross-linking of enzymes with glutaraldehyde involves the reaction between this bifunctional reagent and the residual free amino groups of the enzymes (*Figure 3*). The linkages formed between the amino groups and glutaraldehyde are irreversible and survive extremes of pH and temperature. A procedure describing the cross-linking of glucoamylase immobilised on a hydrous titanium(IV) oxide derivative of controlled pore glass is given in *Table 15*.

The influence of glutaraldehyde concentration, cross-linking time and pH on enzyme activity should be studied, as well as a determination of whether the cross-linked-immobilised enzyme leaks from the support.

The cross-linking step with glutaraldehyde confers better stability on the immobilised enzyme preparations but the increase in stability is obtained at the expense of immobilised enzyme activity.

2.4 Enzyme Immobilisation on Derivatised Transition Metal-activated Supports

Although the metal-link method yields immobilised enzyme preparations with good initial activities, the operational stabilities are frequently poor, mainly with

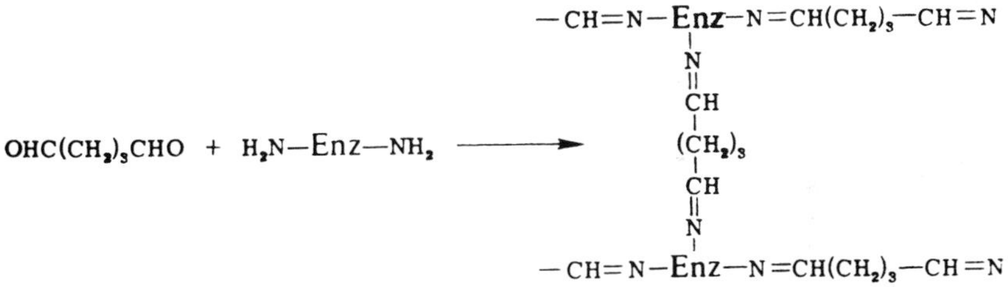

$$OHC(CH_2)_3CHO \ + \ H_2N-Enz-NH_2 \ \longrightarrow$$

$$-CH=N-\underset{\displaystyle \underset{\displaystyle N}{\overset{\displaystyle \|}{\underset{\displaystyle CH}{}}}}{Enz}-N=CH(CH_2)_3-CH=N$$

$$(CH_2)_3$$

$$-CH=N-\underset{\displaystyle \underset{\displaystyle N}{\overset{\displaystyle \|}{\underset{\displaystyle CH}{}}}}{Enz}-N=CH(CH_2)_3-CH=N$$

Figure 3. Cross-linking of enzymes with glutaraldehyde involving the reaction between this reagent and the residual free amino groups of the enzymes.

Table 15. Cross-linking of Glucoamylase Immobilised on Controlled Pore Glass.

1. Immobilise the enzyme on a hydrous titanium(IV) oxide derivative of controlled pore glass as described in Section 2.2.1 (i)
2. Do not wash the immobilised enzyme preparation.
3. Add 20 ml of a 5% w/v glutaraldehyde solution in 0.05 M phosphate buffer, pH 7.0, for 2 h reaction at 4°C.
4. Remove the expent cross-linking agent solution.
5. Wash the cross-linked immobilised enzyme preparation with 20 ml of:
   (i)   0.02 M acetate buffer, pH 4.5, for 10 min;
   (ii)   1 M NaCl in the same buffer for 10 min;
   (iii)   6 M urea for 30 min; and
   (iv)   acetate buffer for 10 min.
6. Keep the immobilised enzyme in the refrigerator until its activity is to be determinined.

macromolecular substrates (6,12). While the stability of the immobilised enzyme preparations increases with the introduction of a cross-linking step, there is a considerable loss of initial activity. Alternative procedures, based on the metal-link method, have been developed in an attempt to produce stable and active immobilised enzyme preparations. These methods usually involve a covalent linkage between the enzyme and the hydrous titanium(IV) oxide derivative of the support. Two procedures based on the diazo linkage and Schiff's base formation are given in *Table 16*. A simple method to immobilise enzymes by covalent attach-

Table 16. Immobilisation of Enzymes on Derivatised Transition Metal-activated Supports.

Diazo coupling

1. Weigh 2 g of porous titanium(IV) oxide and add 40 ml of a 25 mg/ml solution of 1,3-diaminobenzene in 1 M hydrochloric acid. Maintain at 0°C.
2. Add very slowly, with continuous shaking, 30 ml of a 6% aqueous sodium nitrate solution at 0°C.
3. Shake the mixture at 0°C for 30 min and decant.
4. Wash the solid rapidly with 3 x 100 ml of 0.2 M sodium acetate buffer, pH 5.0, at 0°C.
5. Add 20 ml of enzyme solution in the same buffer at 4°C for 2 h.
6. Remove the expent enzyme solution and shake the solid with a saturated solution of 2-naphthol in saturated, aqueous sodium acetate (200 ml) at 4°C for 4 – 5 h.
7. Remove the solution and wash the solid with 5 x 100 ml of 0.2 M sodium acetate buffer, pH 5.0, at 4°C.
8. Store the immobilised enzyme at 4°C in the same buffer.

N.B. Polymerisation of the diazotised 1,3-diaminobenzene molecules occurs prior to coupling, but some amino groups remain unchanged. Subsequent reaction with 2-naphthol is employed to remove any unreacted diazonium groups.

Alkylamine derivatives of transition metal titanium(IV)-activated supports

1. Add 2.5 ml of a 15% w/v titanium(IV) chloride solution in 15% w/v hydrochloric acid to 1 g of controlled pore glass and dry the mixture in an oven at 45°C for 30 h (an oxychloride derivative is obtained).
2. Add 5 ml of a 1% w/v 1,6-diaminohexane solution in carbon tetrachloride to the oxychloride derivative and heat the mixture at 45°C for 30 min (an alkylamine derivative is obtained).
3. Remove the excess amine by decantation and wash with 3 x 10 ml methanol followed by 3 x 10 ml distilled water.
4. Add 5 ml of 5% w/v glutaraldehyde solution in 0.05 M pyrophosphate buffer, pH 8.6, for 1 h at 25°C (an aldehyde derivative is obtained).
5. Remove the excess aldehyde by decantation and wash with 3 x 10 ml distilled water and 10 ml of 0.02 M sodium acetate buffer, pH 4.5.
6. Add 20 ml of the enzyme in 0.02 M sodium acetate buffer, pH 4.5, to the aldehyde derivative for 2 h at 4°C.
7. Wash the immobilised enzyme with 20 ml of each of the following solutions: 0.02 M sodium acetate buffer, pH 4.5 (10 min stirring); 1 M sodium chloride in 0.02 M sodium acetate buffer, pH 4.5 (10 min stirring); 6 M urea (30 min stirring); and 0.02 M sodium acetate buffer, pH 4.5 (10 min stirring).

N.B. The amine solvent is one of the most important steps in the procedure in order to obtain both active and stable immobilised enzyme preparations. The use of water or a hydrophilic solvent for the diamine reagent leads to competition between the solvent and the diamine for the chloride ions on the surface of the oxychloride derivative of the titanium(IV)-activated support, resulting in a lower content of amino groups on the support surface.

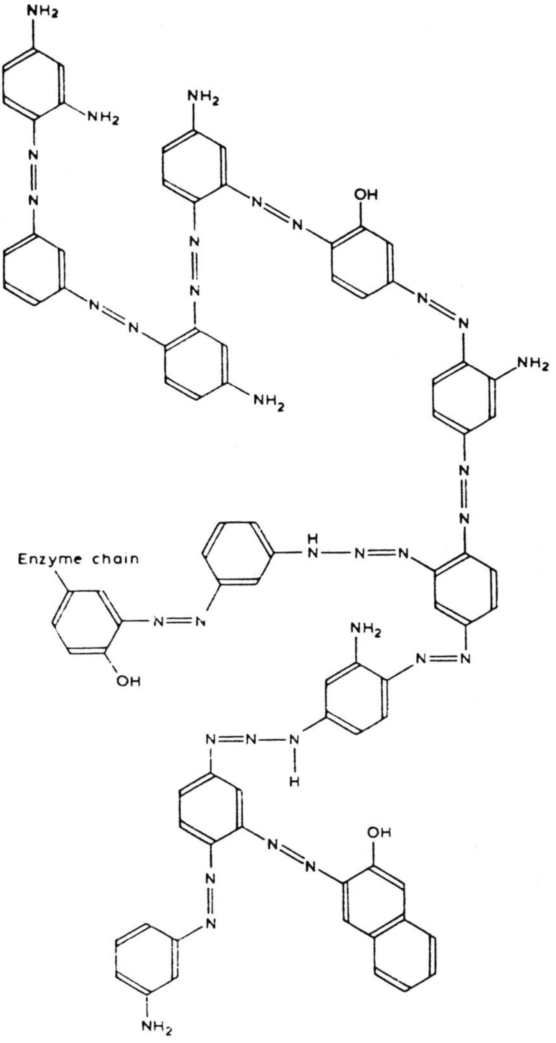

Figure 4. Immobilisation of enzymes by covalent attachment *via* diazo coupling involving the coating of the support with diazotised 1,3-diaminobenzene.

ment *via* diazo coupling involves the coating of the support with diazotised 1,3-diaminobenzene (*Figure 4*).

The Schiff's base formation method is based on the activation of the support with transition metal salts, usually titanium(IV) chloride, followed by amination with a suitable diamine solution in a hydrophobic solvent, in order to obtain an alkylamine derivative, and subsequent reaction with a bifunctional reagent which makes the bridge between the alkylamine derivative of the support and the enzyme.

2.5 Immobilisation of Enzymes and Biomolecules on Hydrous Transition Metal Oxides

A new method of enzyme immobilisation based on the chemistry of the hydrous transition metal oxides, mainly of titanium(IV) and zirconium(IV) has been developed (23 – 25). With this technique the only requirement, besides the enzyme, is one reagent which is commercially available and of adequate stability. A one-step immobilisation process would be more desirable from an economic point of view.

This technique uses the hydrous metal oxides as the only supports (internal) for enzyme immobilisation, as these oxides can be produced by precipitation after the hydrolysis of the corresponding chlorides. The enzyme molecules are immobilised by chelation.

Several methods can be used for the precipitation of hydrous metal oxides with enzymes. The precipitation of the hydrous oxide in the presence of enzyme might yield a more active product than would later addition of the enzyme, owing to the higher surface area of the growing (precipitating) particles. However, to use this kind of co-precipitation, routes have to be chosen to include means of minimising the risks of deactivation of the enzyme through exposure to extreme (acidic) pH values.

In order to obtain the highest bound activities possible, it is necessary to optimise the immobilisation process with respect to some of the more critical parameters, i.e., duration of coupling, pH of coupling and enzyme/hydrous oxide ratio. A procedure using titanium(IV) chloride as metal is described in *Table 17*.

Table 17. Immobilisation of Enzymes on Hydrous Transition Metal Oxides.

1.	Add 20 ml of distilled water to 1 ml (1.73 g) of pure titanium(IV) choride[a].
2.	Immediately neutralise the mixture to pH 7.0 with a 1.0 M ammonia solution.
3.	Centrifuge the resultant suspension and discard the supernatant liquid.
4.	Wash the precipitate thoroughly with 5 x 25 ml distilled water.
5.	Centrifuge and discard the washing supernatant.
6.	Prepare a suspension of hydrous titanium(IV) oxide in an enzyme solution at pH 7.0 and 4°C for 1 h.
7.	Centrifuge the mixture and discard the supernatant liquid.
8.	Wash the precipitate [enzyme-bound hydrous titanium(IV) oxide ⊕ with 10 x 25 ml which contains both 10^{-2} M L-cysteine hydrochloride and 4×10^{-4} M ethylenediamine tetraacetic acid in 0.1 M sodium phosphate buffer, pH 7.0.
9.	Store the immobilised enzyme preparation in 25 ml of the same buffer at 4°C.

[a]Instead of using pure titanium(IV) chloride, an aqueous acidic solution of this salt, such as the commercial solution of 15% w/v of titanium(IV) chloride in 15% w/v hydrochloric acid or a 50% w/v solution of titanium(IV) chloride in 6 M hydrochloric acid can be used. Also, other metal chlorides can be used such as: cobalt(II), copper(II), iron(II), manganese(II), tin(II), zinc(II), chromium(III), iron(III), vanadium(II), tin(IV), and zirconium(IV) chlorides. In a standard method for the preparation of the hydrous metal oxides, solutions are prepared by dissolving the required amount of metal chloride in 1.0 M [5.0 M for tin(IV) chloride ⊕ hydrochloric acid to give a 0.65 M metal solution. Titanium(III) chloride can also be used from its commercial sources as a 12.5% w/v solution in 6 M hydrochloric acid.

3. IMMOBILISATION OF MICROBIAL CELLS BY METAL-LINK/CHELATION PROCESSES

The immobilisation of enzymes by attachment to water-insoluble material has received considerable attention for some time and possible applications have been pursued extensively (2). A logical extension of this approach, especially where multi-stage enzymic reactions are being considered, is the immobilisation of microorganisms which are often the source of many enzyme preparations. The advantages of such an approach are immediately obvious. The tedious and time-consuming procedures for enzyme extraction and purification are instantly eliminated, co-factors and co-enzymes are readily at hand, the cellular enzymes are often organised into the requisite metabolic pathways and problems associated with enzyme instability may also be avoided. Furthermore, the use of immobilised cells would avoid the problem in industrial processes of separating the product from the enzyme.

Investigation of a number of gelatinous hydrous metal oxides (frequently called hydroxides, although their full structures are uncertain) has established that hydrous titanium(IV), zirconium(IV), iron(III), vanadium(III) and tin(II) oxides at least are capable of forming, with enzymes, insoluble complexes which are enzymically active. From the practical viewpoint, hydrous titanium(IV) and zirconium(IV) oxides proved the most satisfactory. Comparatively high retentions of enzyme specific activity may be achieved (24 – 27). Such hydrous metal oxide materials have also proved to be suitable for the immobilisation of amino acids and peptides (24), antibiotics with retention of anti-microbial activity (28), polysaccharides (29), etc.

Hydrous titanium(IV) and zirconium(IV) are insoluble over the normal physiological pH range, and since, when acting as enzyme immobilisation matrices, they give good retention of enzymic activity, they seem to have little or no effect on the function of biologically active molecules. If enzymic activity, which is extremely sensitive to conformational changes in the enzyme molecule, is not seriously affected by immobilisation, there then seems to be little reason why cell walls should be disrupted or destroyed by this process and the cells themselves, therefore, have a good chance of remaining viable.

3.1 Immobilisation of Microbial Cells on Hydrous Metal Oxide

The immobilisation process for the hydrous metal oxides is envisaged as involving the replacement of hydroxyl groups on the surface of the metal hydroxide by suitable ligands from enzyme or cell, resulting in the formation of partial covalent bonds. In the case of enzymes, such ligands could be the side-chain hydroxyls of L-serine or L-threonine, the carboxyls of L-glutamic acid or L-aspartic acid and the ε-amino group of L-lysine residues, oxygen-containing ligands being preferred to those containing nitrogen. In the case of cells, the structural complexity of the cell wall ensures the availability of a great diversity of suitable ligands from both protein and carbohydrate moieties.

An experimental method for microbial cell immobilisation by chelation on hydrous metal oxide is described in *Table 18*.

Table 18. Immobilisation of Microbial Cells on Hydrous Metal Oxide.

1.	To a 15% w/v titanium(IV) chloride solution in 15% w/v hydrochloric acid [or a 0.65 M zirconium(IV) chloride solution in 1.0 M hydrochloric acid ⊕ slowly add a 2.0 M ammonia solution until neutrality (pH 7.0).
2.	Wash the precipitate with 3 x 5.0 ml of a 0.9% w/v saline solution to remove ammonium ions.
3.	To the hydrous metal oxide prepared above add a suspension of *Escherichia coli* cells ($A_{600}^{1.0\ cm}$:0.216) in 10 ml of 0.9% w/v saline. Agitate gently for 5 min at room temperature.
4.	Allow the mixture to stand at room temperature and the suspension to settle out, leaving a clear supernatant ($A_{600}^{1.0\ cm}$:0.222). (Note: this liquid is practically devoid of microorganisms as shown by microscopy).
5.	Consolidate the immobilised cell preparation by centrifugation at low speed and remove the supernatant.

Table 19. Immobilisation of Cells on Transition Metal-activated Supports.

1.	To a 1 g of pumice stone add 3 ml of a 15% w/v titanium(IV) chloride solution in 15% w/v hydrochloric acid.
2.	Dry the mixture in an oven for 48 h at 45°C (an oxychloride derivative is obtained).
3.	Wash the titanium(IV)-activated support with 3 x 10 ml distilled water (a hydrous oxide derivative is obtained).
4.	Add 10 ml of a 2% wet w/v suspension of baker's yeast (*Saccharomyces cerevisiae)* cells in 0.02 M sodium acetate buffer, pH 4.5, for 2 h at 4°C.
5.	Remove the expanded suspension (supernatant) of the immobilised cell preparation.
6.	Wash the immobilised cell preparation with 3 x 10 ml distilled wter and 10 ml acetate buffer.

Immobilised cells of *Saccharomyces cerevisiae* and *Escherichia coli* can be examined for continued viability by measurement of their oxygen uptake (at 25°C in aerated 0.2 M sodium acetate buffer, pH 5.0, or 0.9% w/v saline by use of an oxygen electrode). The rate of oxygen uptake of the immobilised cells was approximately 30% of that of the same number of free cells. Thus, respiration of the cells could continue even when the cells are immobilised. The reduced rate of oxygen uptake is probably caused by the restriction by the metal hydroxide of access of aerated buffer to the cells and a decrease in the cell surface area available for oxygen transfer.

For determining that the cells are firmly attached to the surface of the metal hydroxide and not just loosely trapped in the gelatinous matrix, a different microorganism, *Serratia marcescens*, can be employed. When incubated in nutrient medium at 25°C, this organism produces a distinctive red colouration, which enables the immobilised cells to be distinguished readily from those of any other contaminating microorganisms and also obviates the need for sterile experimental conditions.

On adding cultures of *S. marcescens* to zirconium(IV) and titanium(IV) hydroxides, the red colouration (that is, the cells) became associated with the insoluble matrix, the supernatant and subsequent washing (with saline solution) being almost cell free, thus demonstrating the strength of the cell-metal hydroxide interaction.

An application of transition metal-chelated living cells is the fermentation of acetic acid (manufacture of vinegar) which was reported by Kennedy *et al.* (30).

3.2. Immobilisation of Cells on Transition Metal-activated Supports

Transition metal-activated supports can also be used for cell immobilisation as shown in *Table 19*.

Yeast cells immobilised on a hydrous titanium oxide derivative of pumice stone result in a stable preparation with a high activity.

4. REFERENCES

1. Kennedy,J.F. and Cabral,J.M.S. (1983) in *Solid-Phase Biochemistry: Analytical and Synthetic Aspects,* Scouten,W.H. (ed.), John Wiley, New York, p.253.
2. Kennedy,J.F. and Cabral,J.M.S. (1983) in *Applied Biochemistry and Bioengineering,* Vol. 4, Chibata,I. and Wingard,L.B. (eds.), Academic Press, New York, p. 190.
3. Novais,J.M. (1971) Ph.D. Thesis, University of Birmingham, UK.
4. Barker,S.A., Emery,A.N. and Novais,J.M. (1971) *Process Biochem.,* 6, 11.
5. Emery,A.N., Hough,J.S., Novais,J.M. and Lyons,T.P. (1972) *Chem. Eng.,* 259, 71.
6. Cabral,J.M.S., Cardoso,J.P. and Novais,J.M. (1981) *Enzyme Microb. Technol.,* 3, 41.
7. Gray,C.J., Lee,C.M. and Barker,S.A. (1982) *Enzyme Microb. Technol.,* 4, 143.
8. Rokugawa,K., Fujishima,T., Kurrinaka,A. and Yoshino,H. (1980) *J. Ferment. Technol.,* 58, 509.
9. Kennedy,J.F. and Pike,V.W. (1979) *Enzyme Microb. Technol.,* 1, 31.
10. Kent,C.A. and Emery,A.N. (1974) *J. Appl. Chem. Biotechnol.,* 24, 663.
11. Thornton,D., Flynn,A., Johnson,D.B. and Ryan,P.D. (1975) *Biotechnol. Bioeng.,* 17, 1679.
12. Cardoso,J.P., Chaplin,M.F., Emery,A.N., Kennedy,J.F. and Revel-Chion,L.P. (1978) *J. Appl. Chem. Biotechnol.,* 28, 775.
13. Kennedy,J.F. and Watts,P.M. (1974) *Carbohydr. Res.,* 32, 155.
14. Cabral,J.M.S., Kennedy,J.F. and Novais,J.M. (1982) *Enzyme Microb. Technol.,* 4, 337.
15. Cabral,J.M.S., Kennedy,J.F., Novais,J.M. and Cardoso,J.P. (1984) *Enzyme Microb. Technol.,* 6, 228.
16. Cabral,J.M.S. (1982) Ph.D. Thesis, Technical University of Lisbon, Portugal.
17. Cabral,J.M.S., Novais,J.M. and Cardoso,J.P. (1981) *Biotechnol. Bioeng.,* 23, 2083.
18. Flynn,A. and Johnson,D.B. (1978) *Biotechnol. Bioeng.,* 20, 1445.
19. Kennedy,J.F., Barker,S.A. and White,C.A. (1977) *Starke,* 29, 240.
20. Kennedy,J.F. and White,C.A. (1979) *Starke,* 31, 375.
21. Kennedy,J.F. and Kalogerakis,B. (1980) *Biochimie,* 62, 549.
22. Kennedy,J.F., Kalogerakis,B. and Cabral,J.M.S. (1984) *Enzyme Microb. Technol.,* 6, 68.
23. Kennedy,J.F. and Kay,I.M. (1976) *J. Chem. Soc. Perkin Trans. I.,* 329.
24. Kennedy,J.F., Barker,S.A. and Humphreys,J.D. (1976) *J. Chem. Soc. Perkin Trans. I.,* 962.
25. Kennedy,J.F. and Pike,V.W. (1978) *J. Chem. Soc. Perkin Trans. I.,* 1058.
26. Kennedy,J.F., Humphreys,J.D. and Barker,S.A. (1981) *Enzyme Microb. Technol.,* 3, 129.
27. Kennedy,J.F. and Kay,I.M. (1975) *Carbohydr. Res.,* 44, 291.
28. Kennedy,J.F. and Humphreys,J.D. (1976) *Antimicrob. Agents Chemother.,* 9, 766.
29. Kennedy,J.F., Barker,S.A. and White,C.A. (1977) *Carbohydr. Res.,* 54, 1.
30. Kennedy,J.F., Humphreys,J.D., Barker,S.A. and Greenshields,R.N. (1980) *Enzyme Microb. Technol.,* 2, 209.

CHAPTER 3

Immobilisation of Cells and Enzymes by Gel Entrapment

MAREK P.J.KIERSTAN and MICHAEL P.COUGHLAN

When he nothyng preuailed, he turned to suttle entrappynges

Norton (1)

1. INTRODUCTION

Many thousands of years ago man learned how to make bread and wine and how to tan hides. The ability to provide these basic necessities of every day life did not of course imply an understanding of the biological processes involved. Indeed, only relatively recently have we begun to understand cellular metabolism and the catalytic properties of enzymes. However, acquistion of this knowledge has in turn led to a desire to exploit more fully the potential of cells and enzymes for domestic, industrial and medical purposes.

1.1 Why Immobilisation?

Enzyme specificity is clearly an important consideration when choosing catalysts for reactions of commercial interest. However, the cost of producing the necessary enzymes is often prohibitive. In this context immobilised enzyme preparations may be more effective since they are recoverable and possibly more stable than free enzymes. Indeed several notable examples of the successful application of immobilised enzymes have been documented (see ref. 2 and also below). Nevertheless, it is true to say that immobilised preparations of individual enzymes have not been as successful as originally hoped. Operational instability and the ability to catalyse only a single reaction have frequently been cited as shortcomings. Moreover, the types of enzymes that can be used are 'for practical purposes' restricted to those for which co-factor regeneration is not a requirement. These deficiences rekindled an interest in exploiting the potential of cellular metabolism. Messing (10) rightly pointed out that many examples of immobilised cells are to be found in nature and that one of the first technological applications of such preparations (i.e., the quick vinegar process) took place more than 150 years ago. Many instances of the successful application of bacterial, fungal, higher plant and mammalian cells and subcellular organelles have been reviewed (see refs. 2 – 12). Indeed, several examples are discussed in this volume (Chapters 5 – 9).

The use of immobilised cells obviates the need to isolate and purify the required enzyme(s). Moreover, being in their native environment, operational denaturation of the enzymes should be minimised. Also weighing in favour of cells is the fact that co-factor regeneration can take place under suitable conditions. This extends the range of enzymes that can be exploited and allows the operation of synthetic as well as degradative processes. Moreover, multi-step (i.e., sequential) enzymic reactions are possible. Of course the use of immobilised cells is not without its problems and disadvantages. These include the fact that the cell wall, plasma membrane or subcellular membranes may seriously affect the ability of substrates to get to the appropriate enzymes and of products to diffuse out again. Furthermore, the integrity of cells and the stage of the growth cycle during which the enzyme(s) in question is produced must be maintained. Finally, the multiplicity of enzymes present, which in some instances can prove to be an advantage, can also produce a variety of unwanted side reactions.

1.2 Applications of Immobilised Enzyme or Cells

In general one may consider at least four areas in which immobilised enzyme or cell preparations may find use, i.e., industrial, environmental, analytical and chemotherapeutic. Examples of the first would include amino acid and antibiotic synthesis, steriod transformation, production of sugar syrups and dairy products. Environmental applications include waste water treatment and the hydrolysis of cellulose/hemicellulose/lignin containing wastes of urban, industrial and agricultural origin. Apart from the social desirability of such processes, the generation of usable energy in the form of methane or ethanol may provide the economic justification. In view of their particulate nature the 'activity' of entrapped enzyme or cell preparations cannot always be assayed or exploited directly as one may do with soluble enzymes. However, they are ideally suited to use in 'columns' or continuous-flow stirred reactors through which the substrate to be transformed is passed. The reusability of such columns is a decided advantage as is the ease of obtaining product free of enzyme. Entrapped catalysts may also be used as 'enzyme electrodes' in, for example, measurement of the concentration of glucose or urea in biological or other fluids (see Chapter 5). Mosbach and co-workers (5) developed sensitive enzyme thermistor techniques for measurement of immobilised enzyme activity that obviate the need for optically clear solutions. Microencapsulated enzyme preparations are also being examined for their possible chemotherapeutic use in replacing enzymes that are absent from individuals with certain genetic disorders. Such preparations may also be used as digestive aids, for the removal of toxic metabolites and for prevention of growth of substrate-dependent tumours. Apart from their practical applications it is hoped that studies on entrapped or encapsulated enzymes may elicit a better understanding of their functioning *in vivo*.

2. METHODS OF IMMOBILISATION

2.1 General Principles

Immobilisation may be considered to be the physical separation, during continuous operation, of the catalyst (cell, cell fraction or enzyme) from the solvent in such a way that substrate and product molecules may readily exchange between phases. Separation of the catalyst from the solvent may be achieved by adsorption onto or covalent binding to insoluble organic or inorganic supports (see Chapters 1 and 2). Alternatively, the individual catalysts may be linked together to form aggregates or co-polymers. One disadvantage of such methods of immobilisation is that relatively large amounts of catalyst are required. Moreover, the chemical modification to which the enzymes or cells are subjected during such immobilisation may adversely alter their catalytic and other properties. For these reasons many investigators have, in Norton's words (1), turned to 'suttle entrappynges'. In this technique the cells or enzyme(s) to be used are separated from the bulk phase by physical entrapment or encapsulation. Thus, the catalyst may be entrapped within a polymeric mesh such as polyacrylamide gel or calcium alginate by carrying out the polymerisation and/or cross-linking reaction in the presence of the enzyme(s) or cells in question. Related techniques include encapsulation within liposomes or within nylon or collodion membranes, or physical confinement in ultrafiltration devices. Where appropriate, the catalyst-containing particles are formed into spheres to facilitate practical application.

If the reaction to be catalysed is single step, one may entrap either the relevant enzyme or non-viable cells exhibiting the activity in question. In the former case a high degree of cross-linking of the matrix is essential if one is to maximise entrapment and minimise leakage. However, extensive cross-linking will limit diffusion of substrate and product molecules into and out of the matrix. Such problems may not be encountered with entrapped cell systems. Because of the relatively large size of cells, matrices with a lower degree of cross-linking and, thus, good diffusion properties may be used. Catalysis of multistep reactions, for example, the conversion of glucose to ethanol in which cases co-factor generation and retention is a common requirement, necessitates the entrapment of living cells. In this case one must consider the possible ill effects of the cross-linking agents used on cell viability. Care must also be taken to ensure an adequate supply of oxygen to the cells in question and a means for removal of the carbon dioxide produced as a result of metabolic activity. The difficulties attendant on these apparently simple demands have recently been reviewed (12). Thus, entrapment of viable cells requires that mild immobilisation procedures be used and that the immobilising matrix be an 'open pore system' with good gas transfer properties. These requirements unfortunately tend to yield beads or spheres having low strength – a factor which becomes important especially when processes are scaled up. Despite these limitations, immobilised living cell systems offer an attractive alternative to standard fermentation procedures. Indeed, several investigators claim improvements in productivity of an order of magnitude or more using such alternatives. Moreover, some immobilised cell systems have been found to retain viability for many months.

2.2 Entrapment of Enzymes in Polyacrylamide Gels

Polyacrylamide, the most commonly used matrix for the entrapment of enzymes,

Table 1. Entrapment of Enzymes in Polyacrylamide Gel.

1.	Materials required:
	Enzyme powder or concentrated solution
	0.1 mM EDTA, 0.1 M Tris-HCl, pH 7.0
	0.1 mM EDTA, 0.5 M NaCl, 0.1 M Tris-HCl, pH 7.0
	N,N'-methylene-bis-acrylamide (bisacrylamide)
	Acrylamide
	Dimethylaminopropionitrile
	Potassium persulphate (10 mg/ml); freshly prepared
	A source of nitrogen
	Syringe system (5 or 10 ml capacity fitted with a fine needle ~0.5 mm diameter)
	Vacuum filtration system
	Stoppered flask (50 ml)
2.	Prepare the following solutions:
	Dissolve 110 mg of methylene-bis-acrylamide and 10 mg of acrylamide in 10 ml of 0.1 mM EDTA, 0.1 M Tris-HCl, pH 7.0, and keep in a stoppered bottle. Dissolve the enzyme (usually ~15 mg) in this solution.
3.	Purge with nitrogen for 20 min, this time period is important in order to remove dissolved oxygen which interferes with the polymerisation reaction.
4.	Add 0.2 ml of dimethylaminopropionitrile and mix gently.
5.	Add 0.5 ml of 10 mg/ml potassium persulphate and mix gently. Stopper the flask and allow to stand at room temperature for 20 min. An opaque gel will form.
6.	Disrupt the gel by shaking the flask and then passing the contents through a fine-bore syringe needle.
7.	Wash the gel with 50 ml of 0.1 mM EDTA, 0.5 M NaCl, 0.1 M Tris-HCl, pH 7.0.
8.	Repeat the washing step at least three times or until significant amounts of enzyme activity are no longer detectable in the wash solution.
9.	Resuspend the enzyme-containing gel in 15 ml of 0.1 mM EDTA, 0.1 M Tris-HCl, pH 7.0. A well immobilised system should not leak enzyme into the suspending buffer.

has the property of being non-ionic. This has the effect that the pH profiles characteristic of the free enzyme are minimally altered. Moreover, charged substrates and products do not accumulate nor are they depleted in the matrix. However, the failure of the matrix to interact with the entrapped protein does little to prevent leakage. Generation of a high degree of cross-linking is necessary to obviate this problem. Thus, the polymerisation must be efficient (an oxygen-free medium is the most important factor in this context). Unfortunately, as mentioned in Section 2.1, there are diffusional problems arising from the use of highly cross-linked matrices especially where large substrates are concerned. Thus, while entrapment of enzymes in polyacrylamide gel is commonly used it has certain disadvantages compared with other immobilisation techniques such as covalent binding to inert supports.

A number of procedures for the entrapment of enzymes in polyacrylamide gels have been published. The procedure of Trevan and Grover (13) described in *Table 1* was designed for immobilisation of urease. Thus, one may have to modify the buffers and pH values somewhat to meet the specific requirements of other enzymes.

Table 2. Entrapment of Cells in Cross-linked and Pre-polymerised Linear Polyacrylamides.

1.	Materials required:
	Washed cells
	0.2 M potassium phosphate buffer, pH 7.0
	Acrylamide
	N,N'-methylene-bis-acrylamide (bisacrylamide)
	Tetramethylethylenediamine (TEMED)
	Ammonium persulphate
	Scalpel
	Coarse sieve
	Glass Petri dishes
2.	Suspend about 5 g wet weight of cells in 10 ml of distilled water and chill in ice.
3.	Chill 10 ml of 0.2 M potassium phosphate buffer, pH 7.0, in ice.
4.	Add to the buffer:
	2.85 g acrylamide
	0.15 g bisacrylamide
	10 mg ammonium persulphate. Mix to dissolve these solids.
5.	Immediately mix the chilled buffer solution with the chilled cell suspension, pour into 2 or 3 glass Petri dishes and cover.
6.	Allow polymerisation to proceed for 1 h.
7.	Suspend the sieved gel in 100 ml of 0.2 M potassium phosphate buffer, pH 7.0, allow to settle and then decant the fines.

2.3 Entrapment of Cells in Polyacrylamide Gels

Systems similar to those described in Section 2.2 above can be used for the immobilisation of cells, whether viable or not. However, losses in viability are encountered as a result of the toxicity of the monomers used to form polyacrylamide gel (i.e., acrylamide and bisacrylamide) and the heat evolved during polymerisation. If, on the other hand, the immobilised cells are being used to catalyse a single-step reaction, losses in viability may not be that critical. As discussed in Section 2.1, the large size of cells as compared with enzymes obviates the need for a high degree of cross-linking. For these reasons, techniques have been developed which use cross-linked and pre-polymerised linear polyacrylamides for the entrapment of cells with good retention of viability (see for example ref. 14). One such technique is given in *Table 2*.

2.4 Entrapment of Cells in Calcium Alginate Gels

Alginate, the major structural polysaccharide of marine brown algae, contains β-D-mannopyranosyl uronate and α-L-gulopyranosyl uronate in regular (1-4)-linked sequences (15). Both homopolymeric sequences are found together, although to different extents, in all alginate molecules. Mixed sequences containing both monomers are usually also present. In the presence of monovalent cations, the polysaccharide forms a viscous solution even at low concentrations. In contrast, in the presence of divalent cations, especially calcium, gelation occurs. Since gel formation can take place under mild conditions, entrapment in this matrix is very suitable for immobilisation of viable cells. In fact it is in this context that it has found most extensive application. The procedure described in

Table 3. Entrapment of Cells in Calcium Alginate Gels.

1.	Materials required: Sodium alginate solution (4%, w/v)[a] 0.2 M $CaCl_2$ Cell slurry (~ 10 – 30 g dry weight/100 ml) Syringe (10 ml) fitted with a wide bore needle approximately 1 mm diameter for droplet formation or a 10 ml pipette with an opening of about 3 mm diameter Magnetic stirrer.
2.	Mix equal volumes (50 ml is suitable for preliminary work) of sodium alginate solution and of cell slurry gently together.
3.	Extrude the mixture dropwise *via* a 10 ml syringe from a height of about 20 cm into an excess of 0.2 M $CaCl_2$[b]. One litre of $CaCl_2$ solution is a suitable volume for 100 ml of mixture.
4.	Leave the beads of calcium alginate entrapped cells to harden in the $CaCl_2$ solution for about 20 min.

[a]Preparation of the sodium alginate solution may give problems because the dry powder, being hygroscopic, may form clumps. Accordingly, it is recommended that the powder be added to the water while being stirred and that stirring is continued for a further hour to ensure dissolution. The solution is then left to stand for about 30 min to allow air bubbles to escape. Removal of air bubbles is essential since otherwise they will be entrapped in beads causing these to float and so create problems in use of the beads.
[b]If desired, the gels can be formed into slabs or cubes by pouring the sodium alginate/cell mixture into the appropriate mould and overlaying with a solution of calcium chloride (0.2 – 0.5 M). When formed, the slabs or cubes can be hardened further by incubation with an excess of 0.2 M calcium chloride.

Table 2 is that reported by Kierstan and Bucke (16).

Increasing the concentration of the alginate solution, and to a certain extent, that of the $CaCl_2$ solution will result in more tightly 'cross-linked' gels. It should be noted, however, that high concentrations of alginates are difficult to work with. The concentrations indicated in *Table 3* are satisfactory for most applications. The calcium ions may, if necessary, be replaced by other divalent cations such as barium.

When using calcium alginate systems it is advisable that 10 mM calcium chloride be included in operating streams. It is also important that chelating agents such as phosphates and citrates be avoided as these disrupt the gel structure by binding calcium.

2.5 Entrapment of Cells in ϰ-Carrageenan

Carrageenans are heterogeneous polysaccharides containing predominantly α-D-galactopyranosyl sulphate esters. ϰ-Carrageenan is the insoluble fraction obtained on addition of potassium ions to an aqueous extract of carrageenan. The standard procedure for entrapment in this material is described in *Table 4* and is based on the methods of Tosa *et al.* (17) and Wang and Hettwer (18).

Modifications to this procedure allow the production of cube and membrane forms of carageenan-immobilised cells. For cube preparation, the cell slurry is made up in 2% (w/v) KCl, warmed to 40°C and mixed with two volumes of the carageenan solution also at 40°C. The slab which forms in the appropriate mould is left to harden at 10°C for 30 min and is then soaked in 2% (w/v) KCl at 10°C for 1 h. The resulting gel can then be cut into cubes of the required dimension.

Table 4. Entrapment of Cells in x-Carrageenan[a].

1.	Materials required:
	x-carrageenan (4%, w/v) in 0.9% (w/v) NaCl. Prepare this by heating at 60°C to dissolve the polysaccharide and then maintain the solution at 40°C.
	KCl (2% w/v) kept at room temperature (18 – 20°C).
	Cell slurry (~ 10 – 30 mg dry weight/100 ml).
2.	Mix the warm carrageenan solution with the cell slurry at a ratio of 9:1. While still at 40°C, add the mixture dropwise, using a syringe, to the KCl solution.
3.	Leave the beads so formed to harden in the KCl solution for 20 min.

[a]The extrusion and bead preparation procedures are essentially similar to those for preparation of calcium alginate beads (Section 2.4).

Table 5. Entrapment of Cells in Agar.

1.	Materials required:
	Agar solution: dissolve 100 mg agar in 4.5 ml of 0.9% (w/v) NaCl by heating at 100°C and then cooling to 50°C.
	Cell slurry (10 – 30 mg dry weight/100 ml) suspended in 0.9% (w/v) NaCl solution.
	Nylon (100 mesh) net on a glass plate.
	0.1 M sodium phosphate buffer, pH 7.0.
2.	Add 0.5 ml of the cell slurry to 4.5 ml of the agar solution at 50°C and mix. Cast the mixture onto the nylon net and cool to 5°C. Store the membrane which forms in 0.1 M sodium phosphate buffer, pH 7.0, until required.

Fine films and membranes may be produced by the above procedure if, while still at 40°C, the mixture is poured out to a shallow depth and allowed to cool. The film is then gently overlayered with 2% (w/v) KCl and allowed to harden as before. Clearly the cells used must be stable at 40°C if they are to remain viable during entrapment in carageenan systems.

2.6 Entrapment of Cells in Agar Gels

Agar has been used for the entrapment of cells in the form of spherical beads, blocks and membranes, the latter being the most common. A solution of the immobilising medium is heated at a temperature sufficient to liquify the agar. The cells are then added and are entrapped in the gel which forms on cooling. As for the carageenan system (Section 2.5), the cells must have the ability to withstand the relatively high temperature required. The procedure described in *Table 5* is based on that reported by Matsunaga *et al.* (19) for the entrapment of *Clostridium butyricum* in agar.

2.7 Other Cell Entrapment Procedures

Several other techniques for the entrapment of cells which have been reported are of considerable interest though not in common use. These include entrapment (as opposed to adsorption) in collagen and in cellulose systems (see refs. 20,21).

2.8 Co-entrapment of Enzymes and Cells

A frequently encountered drawback in applications of preparations in which en-

Table 6. Co-entrapment of Yeast and Immobolised β-Glucosidase in Calcium Alginate.

1.	Materials required:
	1 g of swollen CNBr-activated Sepharose-4B (Pharmacia Ltd.)
	2.8 mg of partially purified β-glucosidase[a] dissolved in 5 ml of 0.1 M NaHCO$_3$, 0.5 M NaCl.
	1 M ethanolamine-HCl, pH 8.0
	1 M NaCl, 0.1 M sodium acetate buffer, pH 4.0
	1 M NaCl, 0.1 M borate buffer, pH 8.0
	The reagents needed for the entrapment of cells in calcium alginate gels (*Table 3*).
2.	Immobilise the β-glucosidase using the following method: to 1 g of swollen CNBr-activated Sepharose-4B[b], add 2.8 mg of β-glucosidase in the solution above and gently mix for 2 h.
3.	Separate the support material to which the enzyme is now covalently linked, from the supernatant by decantation or vacuum filtration and incubate in 1 M ethanolamine-HCl, pH 8.0 for 2 h.
4.	Wash the system alternately with 1 M NaCl, 0.1 M sodium acetate, pH 4.0 and with 1 M NaCl, 0.1 M borate buffer, pH 8.0 until the washings are free of enzyme activity.
5.	Prepare a sodium alginate/yeast cell mixture as described in *Table 3*.
6.	Add the immobilised β-glucosidase to this mixture to a maximum of 10%.
7.	Bring about co-entrapment by formation of a calcium alginate gel as described in *Table 3*.

[a]β-glucosidase partially purified from the culture filtrate of *Talaromyces emersonii* (23).
[b]As an alternative to this procedure, Lee and Woodward (25) suggest that the enzyme be immobilised on concanavalin A-Sepharose.

zymes and cells are entrapped together in the same matrix is the loss of enzyme as a result of diffusion. One may overcome this problem by first immobilising the enzyme by covalent attachment to a suitable support. The cells and the immobilised enzyme are then co-entrapped in the matrix of choice. The versatility of co-entrapment of enzymes and cells may be illustrated by the following example. Cellulose, a polymer of glucose, is a vast and renewable potential source of energy. To tap this energy one must first hydrolyse the cellulose to glucose and then ferment the glucose to ethanol. Hydrolysis of the substrate requires the combined actions of endocellulases, exocellulases and β-glucosidases. However, the content of the latter enzyme in culture filtrates of many cellulolytic fungi is low. Therefore, for practical saccharification, such filtrates must be supplemented with β-glucosidase from other sources. One way of doing this would be to set up a system comprised of cellulose, culture filtrates containing cellulases, and calcium alginate gel in which are co-entrapped yeast cells and immobilised β-glucosidase (22 – 24). *Table 6* describes a procedure for the co-entrapment of yeast cells and immobilised β-glucosidase (23). This protocol can be modified to accept other cell-enzyme combinations.

3. REACTOR DESIGN

For all but specialist applications, cells are entrapped in gels in the form of beads rather than as fibres, membranes or blocks. Single-step bioconversions using entrapped enzymes or dead cells have usually been carried out in continuous-flow stirred tanks, fluidised beds or hollow fibre reactors (*Figure 1*). Each of these systems has its limitations in terms of size, pressure drop and unevenness of catalyst distribution. In order to increase surface area and so facilitate diffusion

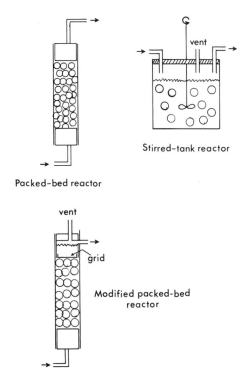

Fig. 1. Reactor types. The stirred-tank reactor may be operated with magnetic or overhead stirrers. The modified packed-bed reactor is for use with gas-producing preparations of immobilised living cells.

(the rate-limiting factor in single-step reactions), it has been the practice to produce small beads of homogeneous size. However, this results in an increase in the pressure drop over the length of the column in use. When immobilised living cells are used, diffusion of gases rather than of other substrates or products is the critical factor. Indeed, gas produced by viable cells may be trapped in the beads causing them to float and thereby decrease their catalytic efficiency. The use of standard packed-bed type reactors should be avoided as gas production may lead to the development of pressures sufficient to burst the reactor vessels. To obviate this danger the reactor should be modified a shown in *Figure 1*. A coarse metal grid (~ 10 mesh) is fitted $2-4$ cm below the outlet port. The grid is either tightly fitted or is held in position by three or four metal extensions which come in contact with the top bung and prevent it from rising. A stirred-tank reactor which can be used in a continuous mode is preferred for routine small-scale investigations. However, the conversion efficiencies of such reactors may be poor, especially when large volumes of substrate are to be processed. One should also keep in mind the possible damage to the beads caused by stirring. In all cases, it is recommended that preliminary studies be carried out with batch-stirrred reactors which can be constructed from simple laboratory equipment.

4. REFERENCES

1. Norton,T. (1561) in *Calvin's Institutions of Christian Religion,* preface.
2. Zaborsky,O.R. (1973) *Immobilised Enzymes,* published by CRC Press, Boca Raton, Florida.
3. Chibata,I. and Tosa,T. (1976) *Appl. Biochem. Bioeng.,* **1,** 329.
4. Abbott,B.J. (1976) *Adv. Appl. Microbiol.,* **20,** 203.
5. Mosbach,K., ed. (1976) *Methods in Enzymology,* Vol. **44,** published by Academic Press, New York and London.
6. Jack,T.R. and Zajic,J.E. (1977) *Adv. Biochem. Eng.,* **5,** 125.
7. Durand,G. and Navarro,J.M. (1978) *Process Biochem.,* **13,** 14.
8. Venkatsubramanian,K., ed. (1979) *Immobilized Microbial Cells,* ACS Symposium Series 106, published by American Chemical Scoiety, Washington, DC.
9. Trevan,M.D. (1980) *Immobilized Enzymes: An Introduction and Applications in Biotechnology,* published by Wiley, New York.
10. Messing,R.A. (1980) *Annu. Rep. Ferment. Processes,* **4,** 105.
11. Brodelius,P. and Mosbach,K. (1982) *Adv. Appl. Microbiol.,* **29,** 1.
12. Mattiasson,B., ed. (1982) *Immobilized Cells and Organelles,* Vols. I and II, CRC Press, Boca Raton, Florida.
13. Trevan, M.D. and Grover, S. (1979) *Trans. Biochem. Soc.,* **7,** 28.
14. Freeman,A. and Aharonowitz,Y. (1981) *Biotechnol. Bioeng.,* **23,** 2747.
15. Rees,D.A., Morris,E.Å., Thom,D. and Madden,J.K. (1982) in *The Polysaccharides,* Vol. **1,** Aspinall,G.O. (ed.), Academic Press, New York and London.
16. Kierstan,M.P.J. and Bucke,C. (1977) *Biotechnol. Bioeng.,* **19,** 387.
17. Tosa,T., Sato,T., Mori,T., Yamamoto,K., Takata,I., Nishida,Y. and Chibata,I. (1979) *Biotechnol. Bioeng.,* **21,** 1697.
18. Wang,H.Y. and Hettwer,D.J. (1982) *Biotechnol. Bioeng.,* **24,** 1827.
19. Matsunaga,T., Karube,I. and Suzuki,S. (1980) *Biotechnol. Bioeng.,* **22,** 2607.
20. Constantinides,A. (1980) *Biotechnol. Bioeng.,* **22,** 119.
21. Linko,Y.-Y., Viskari,R., Pohjola,L. and Linko,P. (1978) *J. Solid-Phase Biochem.,* **2,** 203.
22. Hahn-Hägerdahl,B. (1982) in *Immobilized Cells and Organelles,* Vol. **II,** Mattiasson,B. (ed.), CRC Press, Boca Raton, Florida, p. 79.
23. Kierstan,M.P.J., McHale,A. and Coughlan,M.P. (1982) *Biotechnol. Bioeng.,* **24,** 1461.
24. Coughlan,M.P., McHale,A., Moloney,A.P., O'Rorke,A., Considine,P.J. and Kierstan,M.P.J. (1982) *Trans. Biochem. Soc.,* **10,** 173.
25. Lee,J.M. and Woodward,J. (1983) *Biotechnol. Bioeng.,* **25,** 2441.

CHAPTER 4

Immobilisation of Enzymes by Microencapsulation

HERBERT E.KLEI, DONALD W.SUNDSTROM
and DONGSEOK SHIM

1. INTRODUCTION

In contrast to many of the other methods of enzyme immobilisation presented in other chapters, microencapsulation focuses upon maintaining the solution environment around the enzyme rather than upon maintaining the physical or chemical forces necessary for immobilisation. The original solution containing the enzyme is wholly immobilised rather than selectively immobilising the particular enzyme molecule. Microencapsulation creates artificial cells which have a membrane similar to natural cells to control the size of molecules which can transport into or out of the cell. Large molecules such as enzymes or proteins can be retained within the encapsulated sphere, while small substrate and product molecules can freely diffuse across the synthetic membrane. One of the advantages of microencapsulation over regular entrapment of enzymes is the high surface area possible per unit of enzyme immobilised, allowing high effectiveness factors and high concentrations of enzymes in the original solution. Since the membrane-forming process does not depend upon the enzyme in solution, a large variety of enzymes, cells or biomolecules could be simultaneously encapsulated allowing multi-step reactions to proceed.

Although the concept of forming a polymeric membrane around an enzyme solution to make a microsphere is rather appealing, in practice a great deal of technology is needed to make the spheres uniform and to impart high activity retention. Even though microencapsulated products are currently on the market, they usually have organic liquid interiors containing oils or perfumes, or are aqueous but without catalytic activity. A great deal of current research is directed towards controlled drug release from microcapsules (1,2). The diameter of the microspheres can range from several microns to several thousand microns, and the membrane thickness can go from hundreds of Angstroms to several microns (1,3). It is possible to microencapsulate combinations of enzymes, co-factors and their enzymes for regeneration, proteins, whole cells, ion-exchange resins, activated carbon particles and magnetic particles. Even small microcapsules can be trapped within larger microcapsules as shown in *Figure 1* where small spheres containing catalase are contained within a larger sphere bounded by a butyl rubber membrane.

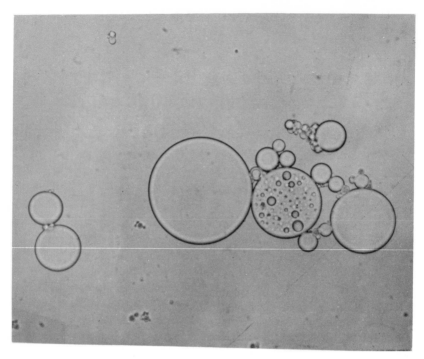

Figure 1. Microencapsulated catalase with butyl rubber membrane showing small capsules within larger one.

2. METHODS OF MICROENCAPSULATION

2.1 Applications to Enzymes and Biochemical Systems

A variety of enzymes and co-factors have been microencapsulated within polymer membranes and still have activity on substrates which diffuse into the capsule. The retained activity is, in general, remarkably lower than for the free enzyme, as seen in *Table 1*. During the microencapsulation, the enzyme in the aqueous microsphere comes into contact with organic solvents and chemical monomers which can readily denature it. In order to protect the enzyme from deactivation, it is usually mixed with a polymer such as bovine serum albumin (BSA) (12), haemoglobin (1), or polyethyleneimine (PEI) (10) prior to microencapsulation. The retained activity of arginase has been investigated for several water-soluble polymers including BSA, polyvinyl pyrolidone (PVP), polyvinyl alcohol (PVA), polyethylene glycol (PEG), dextran and sodium heparinate (13). In general, BSA, PVP, PVA and PEG increase the retained activity of the arginase while the dextran and sodium heparinate have no influence. The hydrophobic portions of the first group can coat the enzyme and protect it during encapsulation while the latter two being strongly hydrophilic would not offer protection. The concentration of the polymer is also important in activity retention. Arginase activity retention was shown to reach a maximum at 1.0% w/v BSA (12). The concentration of protein or other polymer is also important in maintaining the osmotic pressure.

Table 1. Microencapsulated Enzymes and Co-factors.

Enzyme	Membrane	% Activity retention after microencapsulation	Half life	Temperature °C	Reference
Urease	Cellulose acetate butyrate	20	>1 month	25	4
	Cellulose nitrate	37	1 week	37	5,6
	Lipid-polyamide				7
β-Galactosidase	Cellulose nitrate	99			8
Catalase	Cellulose nitrate	99	1.3 day	37	1,9
NADP⁺ ⎱	Lipid-polyamide				7
NAD⁺ ⎰	PEI-nylon		10 days	4	10
Dehydrogenase	PEI-nylon		10 days	4	10
Arginase	Nylon	32	>16 days	37	11
	Poly(phthaloyl piperazine)	12	>60 days	4	12
	Epoxy	14			1
Chymotrypsin	PEI-nylon				1
Asparaginase	Cellulose nitrate	99	5 days	37	1

The concentration of the emulsifier used to form the microspheres in the organic solvent also has a marked effect on activity retention. The activity of microencapsulated arginase rapidly drops to zero as the percentage of Span 85 is lowered from 0.1 to 0.01% v/v (12). Above 0.1% v/v the activity remains constant indicating that a critical concentration is necessary to cover the surface of the water droplets, thus preventing the enzyme from coming into direct contact with the organic solvent. Co-factor recycling within the microcapsule has been achieved by attaching it to a dextran prior to encapsulation, thus making the co-factor too big to diffuse through the membrane (7,10).

The Michaelis-Menten constant usually increases upon microencapsulation. Free urease with a K_m of 3 mM has a K_m of 8 mM after microencapsulation and this value increases to 0.14 M after storage for 16 days at 4°C (6). The K_m for arginase rises by about 50% after microencapsulation (11). In general higher K_m values usually signify increased intraparticle mass transfer resistances resulting either from the membrane or from boundary layers around the particle (14).

2.2. Experimental Procedures

The physical and chemical properties of the microencapsulated enzyme depend upon the nature of the coating material. The polymer should be cohesive to form a continuous film, have permeability to the substrate and be non-reactive in the reaction media. Coating materials are selected from natural and synthetic polymers. Some common ones are carboxymethyl cellulose, nylon, cellulose nitrate, gelatin, silicone rubber, cellulose acetate phthalate, epoxy resins, polyurethanes, gum arabic, styrene and polyamides.

Although several microencapsulation techniques are described in the literature

(1), coacervation and interfacial polymerisation are most widely used for enzyme systems. Coacervation involves a phase separation of colloidal particles of polymer which associate around a small aqueous droplet and then form a continuous membrane by the coalescence of the droplets. Since there are no chemical reactions in coacervation, it has wide applicability for many biochemicals and can produce very thin membranes (1). Typical polymers are cellulose nitrate, cellulose acetate and butyl rubber. Microencapsulation involves the following steps.

(i)   Dissolve the enzyme in a buffer solution usually containing another protein, such as albumin, for protection.

(ii)   Separately prepare an organic phase containing a small amount of water-oil emulsifying agent, such as Span 85. The organic phase should be immiscible with water and common systems are ether, cyclohexane and toluene.

(iii)   Add the aqueous enzyme solution to the organic phase and stir for a predetermined time and stirrer speed. The speed of the stirrer determines the size of the microcapsules.

(iv)   To this two-phase mixture, add a second organic solution. This solution contains the polymer and an organic solvent for the polymer in more of the organic phase originally selected to contain the Span 85. When the second organic liquid is added, small colloids are precipitated out by having the organic solvent diluted.

(v)   As the solvent evaporates out of the organic phase, further precipitation will result with a thicker membrane being formed. The thickness of the membrane is controlled by the amount of polymer added to the organic phase and by the time allowed for phase precipitation.

(vi)   Separate the small microcapsules from the organic phase either by sedimentation or centrifugation and wash with buffered aqueous solution containing Tween 20.

Concentrations and times are critical for effective microencapsulation and suitable values can be found in the literature (1,2,8,15). A preparation procedure for the microencapsulation of catalase is given in *Table 2*.

The interfacial polymerisation procedures places one reactant in the aqueous phase and the other in the organic phase. The polymerisation reaction occurs at the interface and a polymer layer is developed around the aqueous droplet in the organic phase. A common interfacial reaction is the formation of nylon. The aqueous system containing the enzyme, buffer and albumin now also contains a diamine such as 1,6 hexamethylene diamine. The organic solution contains the Span 85 and a mixture of chloroform and cyclohexane. After the emulsion is formed, sebacyl chloride in a chloroform/cyclohexane mixture is added to the emulsion with mixing. After mixing for several minutes, the cells are separated by centrifugation and washed with an aqueous Tween 20 solution. The limitation with the polymerisation method is the possible reaction between the diamine and the enzyme in the aqueous phase. Again, concentrations and times are important in obtaining active immobilised enzymes and can be obtained from published procedures (15,16). Epoxy resins and polyurethane films have also been formed by interfacial polymerisation.

Table 2. Microencapsulation Procedure for Catalase by Butyl Rubber.

1. Prepare the following four mixtures:

    Mixture A: 2.60 g Span 85[a]
    25 ml cyclohexane

    Mixture B: 1.0 ml of aqueous catalase solution in 50 mM phosphate buffer, pH 7.0 at 650 units/ml

    Mixture C: 0.1 g of butyl rubber
    50 ml of cyclohexane
    stir for 1 – 2 h for complete solution

    Mixture D: 15 ml cyclohexane
    10 ml vegetable oil

2. Add mixture A to mixture B with continuous stirring.

3. Follow by the addition of mixture C with continuous stirring.

4. Add mixture D with stirring.

5. Draw a vacuum of 26 mm Hg with continuous stirring to evaporate the cyclohexane solvent. Evaporation must not be too rapid to avoid deposition of polymer particles instead of aqueous microspheres.

6. Centrifuge oil phase containing the microcapsules and pour off the top oil layer. Add 15 ml of phosphate buffer, resuspend the particles and re-centrifuge. Pour off aqueous layer and repeat several times.

7. Resuspend the microcapsules[b] in a buffer solution containing Tween 20 to keep the particles separated.

[a]Span 85 may be obtained from Imperial Chemical Industries (UK or USA).
[b]Microcapsules of catalase may be assayed by adding a known volume of the suspension to a solution of hydrogen peroxide whose disappearance is monitored.

3. MASS TRANSFER CONSIDERATIONS

The practical use of microencapsulated enzymes may be controlled by rates which are limited by mass transfer resistances around the membrane. The diffusion rate across the membrane, as well as diffusion rates across stagnant boundary layers inside of, and external to, the microcapsule, are often the limiting resistances. The importance of the diffusion rate across the membrane is reflected in the increase in K_m of the enzyme upon microencapsulation. An increase in the substrate concentration is necessary to overcome this resistance in a continuous flow stirred reactor (17). With encapsulated β-galactosidase, effectiveness factors were measured between 0.3 and 0.7 and were correlated with the diffusion properties of the membrane (16). The mass transfer rate across the external film was found controlling in the release of microencapsulated materials as evidenced by its dependence on the rate of agitation (18) and has been correlated by an external diffusion model (19). It is possible to have too much enzyme within the microcapsule so that the substrate cannot diffuse across the external boundary layer fast enough to satisfy the maximum possible reaction rates of the encapsulated enzyme. Carbonic anhydrase has been encapsulated with such high reactivity that effectiveness factors of only 0.003 were measured (20). Mixing within the microcapsule can also be important and on several occasions the rate of transport changes with the square root of the reaction time, according to a penetration theory model (21). All of the mass transfer resistances have been combined into a unified model for microcapsules and applied to microencapsulated urease (22). In general this unified model requires fitting several parameters to experimental data, and for Michaelis-Menten kinetics requires trial and error computer solutions.

4. REFERENCES

1. Chang,T.M.S. (1977) in *Biomedical Applications of Immobilized Enzymes and Proteins,* Chang,T. (ed.), Plenum Press, New York, p. 69.
2. Patwardhan,S.A. and Das,K. (1983) in *Controlled-Release Technology, Bioengineering Aspects,* Das,K. (ed.), Wiley and Sons, New York, p. 122.
3. Madan,P.L. (1979) in *Microencapsulation – New Techniques and Applications,* Kondo,T. (ed.), Techno, Inc., Tokyo, p. 11.
4. Gardner,D.L. and Emmerling,D.C. (1977) in *Biomedical Applications of Immobilized Enzymes and Proteins,* Chang,T. (ed.), Plenum Press, New York, p. 163.
5. Chang,T.M.S. (1977) in *Biomedical Applications of Immobilized Enzymes and Proteins,* Chang,T. (ed.), Plenum Press, New York, p. 147.
6. Artmanis,A., Neufeld,R.J. and Chang,T.M.S. (1984) *Enzyme Microb. Technol.,* **6,** 135.
7. Yu.T.T. and Chang,M.S. (1982) *Enzyme Microb. Technol.,* **4,** 327.
8. Wadiak,D.T. and Carbonell,R.G. (1975) *Biotechnol. Bioeng.,* **17,** 1157.
9. Magensen,A. and Vieth,W. (1973) *Biotechnol. Bioeng.,* **15,** 467.
10. Grunwald,J. and Chang,T.M.S. (1981) *J. Mol. Catal.,* **11,** 83.
11. O'Grady,P. and Joyce,P. (1981) *Enzyme, Microb. Technol.,* **3,** 149.
12. Kondo,T. and Muramatsu,N. (1976) in *Microencapsulation,* Nixon,J.R. (ed.), Marcel Dekker,Inc., New York, p. 67.
13. Yanagibashi,N., Arakawa,M., and Kondo,T. (1979) in *Microencapsulation,* Kondo,T. (ed.), Techno, Inc., Tokyo, p. 325.
14. Moo-Young,M., and Kabayashi,T. (1972) *Can. J. Chem. Eng.,* **50,** 162.
15. Chang,T.M.S., MacIntosh,F.C. and Mason,S.G. (1976) *Can J. Physiol. Pharmacol.,* **44,** 115.
16. Wadiak,D.T., and Carbonell,R.G. (1975) *Biotechnol. Bioeng.,* **17,** 1157.
17. Artmanis,A., Neufeld,R.J. and Chang,T.M.S. (1984) *Enzyme Microb. Technol.,* **6,** 135.
18. Nang,L.S. and Carlier,P.F. (1976) in *Microencapsulation,* Nixon,J.R. (ed.), Marcel Dekker Inc., New York, p.185.
19. Takenaka,H., Kawashima,N., Saitoh,M. and Ishibashi,R. (1979) in *Microencapsulation,* Tamotsu,K., (ed.), Techno, Inc., Tokyo, p.35.
20. Dean,D.N., Fuchs,M.J., Schaffer,J.M. and Carbonell,R.G. (1977) *Ind. Eng. Chem. Fund.,* **16,** 452.
21. Bennett,C.O. and Myers,J.E. (1974) *Momentum, Heat and Mass Transfer,* published by McGraw-Hill, New York.
22. Vieth,W.R., Mendiratta,A.K., Mogensen,A.O., Saini,R., Venkatasubramanian,K. (1973) *Chem. Eng. Sci.,* **28,** 1013.

CHAPTER 5

Immobilised Enzyme Electrodes

GEORGE G. GUILBAULT and GRACILIANO de OLIVERA NETO

1. INTRODUCTION

The enzyme electrode is a combination of an ion-selective electrode base sensor with an immobilised (insolubilised) enzyme, which provides a highly selective and sensitive method for the determination of a given substrate.

Clark and Lyon (1) first introduced the concept of the 'soluble' enzyme electrode, but the first working electrode was reported by Updike and Hicks (2) using glucose oxidase immobilised in a gel over a polarographic oxygen electrode to measure the concentration of glucose in biological solutions and tissues. These were both voltammetric or amperometric probes, i.e., the current produced upon application of a constant applied voltage is measured. The first potentiometric (no applied voltage, the voltage produced is monitored) enzyme electrode was described by Guilbault and Montalvo for urea in 1969 (3). Since this time, over 100 different electrodes have appeared in the literature; a summary of these can be found in reference 4 and in *Table 1*.

In enzyme electrodes, the enzyme is usually immobilised, this reduces the amount of material required to perform a routine analysis, and eliminates the need for frequent assay of the enzyme preparation in order to obtain reproducible results. Furthermore, the stability of the enzyme is often improved when it is incorporated in a suitable gel matrix. For example, an electrode for the determination of glucose prepared by covering a platinum electrode with chemically bound glucose oxidase has been used for over 300 days (5).

Of the two methods used to immobilise an enzyme; (i) the chemical modification of the molecules by the introduction of insolubilising groups, and (ii) the physical entrapment of the enzyme in an inert matrix, such as starch or polyacrylamide, the technique of chemical immobilisation is the best to make electrode probes.

2. OPERATIONAL CHARACTERISTICS OF ELECTRODES

An enzyme electrode operates *via* a five-step process:

(i)   the substrate must be transported to the surface of the electrode;
(ii) the substrate must diffuse through the membrane to the active site;
(iii) reaction occurs at the active site;
(iv) the product formed in the enzymatic reaction is transported through the membrane to the surface of the electrode;
(v)   product is measured at the electrode surface.

Table 1. Typical Electrodes and Their Characteristics.

Type	Enzyme	Sensor	Immobilisation[a]	Stability	Response time	Amount of enzyme (U)	Range (mol/l)[b]
1. Urea	Urease (EC 3.5.1.5)	Cation	Physical	3 weeks	30–60 sec	25	$10^{-2}-5 \times 10^{-5}$
		Cation	Physical	2 weeks	1–2 min	75	$10^{-2}-10^{-4}$
		Cation	Chemical	>4 months	1–2 min	10	$10^{-2}-10^{-4}$
		pH	Physical	3 weeks	5–10 min	100	$5 \times 10^{-2}-5 \times 10^{-5}$
		Gas (NH$_3$)	Chemical	4 months	2–4 min	10	$5 \times 10^{-2}-5 \times 10^{-5}$
		Gas (NH$_3$)	Chemical	20 days	1–4 min	0.5	$10^{-2}-10^{-4}$
		Gas (CO$_2$)	Physical	3 weeks	1–2 min	25	$10^{-2}-10^{-4}$
2. Glucose	Glucose oxidase (EC 1.1.3.4)	pH	Soluble	1 week	5–10 min	100	$10^{-1}-10^{-3}$
		Pt(H$_2$O$_2$)	Physical	6 months	12 sec kinetic[c]	10	$2 \times 10^{-2}-10^{-4}$
		Pt(H$_2$O$_2$)	Chemical	>14 months	1 min steady-state	10	$2 \times 10^{-2}-10^{-4}$
		Pt(H$_2$O$_2$)	Soluble		1–2 min	10	$10^{-2}-10^{-4}$
		Pt(quinone)	Soluble	<1 week[d]	3–10 min	10	$2 \times 10^{-2}-10^{-3}$
		Pt(O$_2$)	Chemical	>4 months	1 min	10	$10^{-1}-10^{-5}$
		I$^-$	Chemical	>1 month	2–8 min	10	$10^{-3}-10^{-4}$
		Gas (O$_2$)	Physical	3 weeks	2–5 min	20	$10^{-2}-10^{-4}$
	Glucose oxidase (EC 1.1.3.4) and Catalase (EC 1.11.1.6)	Gas (O$_2$)	Chemical	>3 weeks	2–5 min	10	$2 \times 10^{-2}-10^{-4}$
3. L-Amino acids (general)[e]	L-amino acid oxidase (EC 1.4.3.2)	Pt(H$_2$O$_2$)	Chemical	4–6 months	12 sec kinetic[c]	10	$10^{-3}-10^{-5}$
		Gas (O$_2$)	Chemical		2 min	10	$10^{-2}-10^{-4}$
		Pt(O$_2$)	Chemical	>4 months	1 min	10	$10^{-2}-10^{-4}$
		Cation	Physical	2 weeks	1–2 min	10	$10^{-2}-10^{-4}$
		NH$_4^+$	Chemical	>1 month	1–3 min	10	$10^{-2}-10^{-4}$
		I$^-$	Chemical	>1 month	1–3 min	10	$10^{-3}-10^{-4}$

Table 1 continued.

Type	Enzyme	Sensor	Immobilisation[a]	Stability	Response time	Amount of enzyme (U)	Range (mol/l)[b]
L-Tyrosine	L-Tyrosine decarboxylase (EC 1.1.25)	Gas (CO_2)	Physical	3 weeks	1 – 2 min	25	$10^{-1} - 10^{-4}$
L-Glutamine	Glutaminase (EC 3.5.1.2)	Cation	Soluble	2 days[d]	1 min	50	$10^{-1} - 10^{-4}$
L-Glutamic acid	Glutamate dehydrogenase (EC 1.4.1.3)	Cation	Soluble	2 days[d]	1 min	50	$10^{-1} - 10^{-4}$
L-Asparagine	Asparaginase (EC 3.5.1.1)	Cation	Physical	1 month	1 min	50	$10^{-2} - 5 \times 10^{-5}$
4. D-Amino acids (General)[f]	D-amino acid oxidase (EC 1.4.3.3)	Cation	Physical	1 month	1 min	50	$10^{-2} - 5 \times 10^{-5}$
5. Lactic acid	Lactate dehydrogenase (EC 1.1.1.27)	Pt($[Fe(CN)_6]^{4-}$]	Soluble	<1 week	3 – 10 min	2	$2 \times 10^{-3} - 10^{-4}$
6. Succinic acid	Succinate dehydrogenase (EC 1.3.99.1)	Pt(O_2)	Physical	1 week	1 min	10	$10^{-2} - 10^{-4}$
7. Acetic, formic acids	Alcohol oxidase (EC 1.1.3.13)	Pt(O_2)	Chemical	>4 months	30 sec	10	$10^{-1} - 10^{-4}$
8. Alcohols[g]	Alcohol oxidase (EC 1.1.3.13)	Pt(H_2O_2)	Soluble	1 week	12 sec kinetic[c]	10	0.5 – 100 mg%
		Pt(H_2O_2)	Soluble	1 day[d]	1 min	1	0.5 – 50 mg%
		Pt(O_2)	Chemical	>4 months	30 sec	10	0.5 – 100 mg/%
9. Penicillin	Penicillinase (EC 3.5.2.6)	pH	Physical	1 – 2 weeks	0.5 – 2 min	400	$10^{-2} - 10^{-4}$
			Soluble	3 weeks	2 min	1000	$10^{-2} - 10^{-4}$

Table 1 continued.

Type	Enzyme	Sensor	Immobilisation[a]	Stability	Response time	Amount of enzyme (U)	Range (mol/l)[b]
10. Uric acid	Uricase (EC 1.7.3.3)	$Pt(O_2)$	Chemical	4 months	30 sec	10	$10^{-2}-10^{-4}$
11. Amygdalin	β-Glucosidase (EC 3.2.1.21)	CN^-	Physical	3 days[g]	10 – 20 min	100	$10^{-2}-10^{-5}$
12. Cholesterol	Cholesterol oxidase (EC 1.1.3.7)	$Pt(H_2O_2)$	Soluble		2 min		$10^{-2}-10^{-4}$
13. Phosphatase	Phosphatase/ glucose oxidase (EC 3.1.3.1/ 1.1.3.4)	$Pt(O_2)$	Chemical	4 months	1 min	10 each	$10^{-2}-10^{-4}$
14. Nitrate	Nitrate reductase/ nitrate reductase (EC 1.9.6.1/ 1.6.6.4)	NH_4^+	Soluble		2 – 3 min	10	$10^{-2}-10^{-4}$
15. Nitrite	Nitrate reductase (EC 1.6.6.4)	NH_3 (gas)	Chemical	3 – 4 months	2 – 3 min	10	$5 \times 10^{-2}-10^{-4}$
16. Sulphate	Aryl sulphatase (EC 3.1.6.1)	Pt	Chemical	1 month	1 min	10	$10^{-1}-10^{-4}$

[a] 'Physical' refers to polyacrylamide gel entrapment in all cases; 'chemical' is attachment chemically to glutaraldehyde with albumin, to polyacrylic acid, or to acrylamide, followed by physical entrapment.
[b] Analytically useful range, either linear or with reasonable change if curvature is observed.
[c] 'Kinetic', rate of change in current measured after 12 sec; 'steady-state', current reaches a maximum in 1 min.
[d] Preparation lacks stability as evidenced by constant decrease in signal each day.
[e] Electrode responds to L-cysteine, L-leucine, L-tyrosine, L-tryptophan, L-phenylalanine, and L-methionine.
[f] Electrode responds to D-phenylalanine, D-alanine, D-valine, D-methionine, D-leucine, D-norleucine, and D-isoleucine.
[g] Time required for signal to return to base line before re-use.

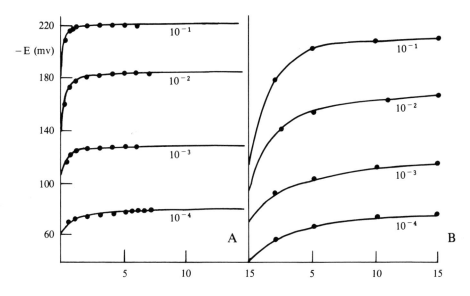

Figure 1. Amygdalin response-time curves for an electrode containing 1 mg of β-glucosidase immobilised by a dialysis paper. (**A**) At pH 7.0; (**B**) at pH 10.0. Reprinted with permission from reference 6.

The first step, transport of the substrate, is most critically dependent on the stirring rate of the solution, so that rapid stirring will bring the substrate very rapidly to the electrode surface. If the membrane is kept very thin, using highly active enzyme, then steps (i) and (iv) are eliminated or minimised; since step (iii) is very fast, the response of an enzyme electrode should theoretically approach the response time of the base sensor. Many researchers have shown with experimental data that one can approach this behaviour by using a thin membrane and rapid stirring. For example, *Figure 1* shows a comparison of the characteristics of the amygdalin electrode using β-glucosidase; on the left-hand side are data obtained by Mascini and Liberti (6) at the optimum pH for the cyanide sensor, pH 9.0. Realising that there is no free cyanide except at very high pH, say 9.0 or 10.0, Rechnitz reasoned that this optimum pH should be used in order to get the best sensitivity. However, at this high pH the enzyme reaction is killed and the rate of conversion becomes very slow. If the optimum pH of the enzyme reaction is used, even though there is very little cyanide at this pH, the response is almost instantaneous, since the enzyme layer is in such intimate contact with the base probe. This has been shown in many cases; for example, Anfalt *et al.* (7) in Sweden showed with the urea-urease reaction that the ammonia liberated from this reaction could be measured more effectively at pH 7.0 or 7.5 than at pH 9.0 or 10.0, because the enzyme reaction was functioning much better at this low pH. Another factor often overlooked in the use of ion-selective electrodes is that the stirring rate not only will promote a faster response at the enzyme electrode or at any probe but also will affect the equilibrium potential or the equilibrium pH that is measured. This becomes very critical: if one is going to stir, one has to stir at a constant rate; otherwise, a different potential value will be obtained every time the assay is performed.

The stability of the electrode depends on the type of entrapment. Here again there is much ambiguous reporting of immobilisation data in the literature. Some researchers use dry storage for a long period of time and then report a fantastically long lifetime. Realistically, the immobilisation characteristics and the stability of the enzyme should be defined both in terms of dry storage and use storage. The lifetime of most soluble enzymes, except perhaps some types of glucose oxidase which are quite stable in the crude form, is generally about 1 week or 25−50 assays. (However, one must realise that there are potential problems in the use of soluble enzymes which are not found in the use of an entrapped enzyme.) The physically entrapped enzyme lasts about 3−4 weeks or 50−200 assays. For the chemically bound enzyme, 200−1000 assays is a good range. In many cases, we and others have achieved at least this. Furthermore there are many immobilised enzymes available, bound onto nylon tubes (such as those Technicon is producing for use on SMAC or the Auto Analyser, those Boehringer has been experimenting with, and those Miles is selling under the trade name Catalink), which are very stable. These tubes have been used for more than 10 000 assays.

Stability is also dependent on the content of enzyme in the gel, on the optimum conditions, as was mentioned, and on the stability of the base sensor itself. *Figure 2* shows the stability of some electrodes using immobilised glucose oxidase prepared by various methods: physical entrapment in a gel and covalent bonding using glutaraldehyde or diazo coupling. The type of chemical bonding serves two purposes:

(i)   it selects the pH range,

(ii) it provides the best immobilisation method for each enzyme. This is shown below for some studies of cholesterol.

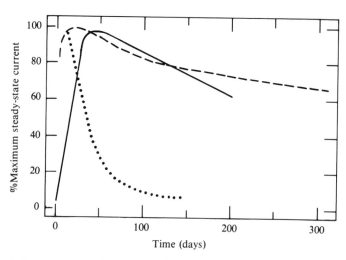

Figure 2. Long-term stability of glucose electrodes by the steady-state method. Glucose oxidase immobilised by: physical entrapment (**A**), covalent coupling using glutaraldehyde (**B**) and diazo coupling (**C**).

Interferences can be in the sensor itself or from other substrates for the enzyme. For example, alcohol oxidase can be used for an excellent acetic acid electrode; an interference would be the native substrate, ethanol. Inhibitors of the enzyme are also interferences, and here immobilising an enzyme makes it very susceptible to environmental factors.

3. EXPERIMENTAL

3.1 Apparatus

The enzyme electrode, once structured, is used like any other ion-selective electrode. The typical apparatus used is shown in *Figure 3*.

The potentiometric probes, e.g., urea, amino acids, penicillin, are plugged directly into a digital voltmeter (e.g., Orion, Corning, Sargent, etc.). The mV read for each concentration tested are then plotted *versus* concentration in a linear-log plot.

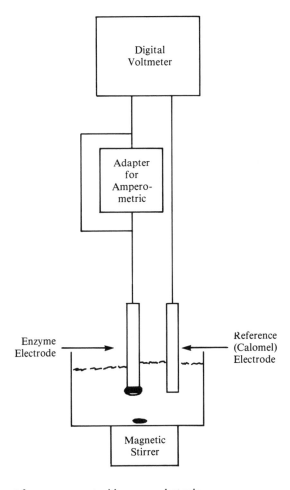

Figure 3. Equipment for measurement with enzyme electrode.

To use the amperometric electrodes, e.g., alcohol or glucose (which are based on use of a Pt or O_2 electrode) a polarographic apparatus, like the Princeton Applied Research Model 170 or 174, can be used. A potential of either $\pm$ 0.85 V is applied to the electrode, and the current (μA) is recorded. A plot of current (μA) *versus* concentration (M) is a linear-linear plot.

To eliminate the necessity of using polarography to monitor an amperometric electrode, Universal Sensors, Inc. (PO Box 736, New Orleans, LA 70148), markets an electrode adapter, which is a device that will simultaneously apply a potential to the amperometric glucose or alcohol probe, take the resulting current generated, and convert it to a voltage which can then be read directly on any voltmeter (Orion, Corning, etc.), provided the scale is $\pm$ 1.7 V. This simplifies the operational use of all enzyme electrodes, to the requirement of only a voltameter plus the adaptor (the latter sells for only about $125).

A reference electrode, generally a calomel electrode, is used together with the enzyme electrodes. Alternatively, the reference electrode can be combined as an integral part of the enzyme electrode as is the case with the Universal Sensors NH_3, CO_2 or O_2 electrode base sensors used in the urea, amion acid, glucose or alcohol probes.

Finally, the electrode must be kept in a solution with *constant* stirring rate, since it has been shown (see reference 8) that a change in stirring rate will change the potential of the electrode measured. This factor was discussed in Section 2.

3.2 Material Required

To construct the enzyme electrode (see Section 3.3) one must first select the appropriate enzyme. From information in standard reference books on enzymology, such as Biochemists Handbook, find an enzyme system which is suitable for your determination. In the ideal case, this will involve the use of the primary function of the enzyme, i.e., the main substrate-enzyme reaction. For example, to construct a glucose electrode, use glucose oxidase, for a urea electrode, use urease. In some cases, you may use an enzyme that acts on the compound of interest as a secondary substrate, e.g., alcohol oxidase for an acetic acid electrode.

Next, obtain the enzyme. Having found the enzyme to be used for your application, check the catalogs of commercial suppliers: Sigma, Boehringer, Worthington, Calbiochem, Miles, etc., to see if the enzyme can be purchased, and to its purity. Pay special attention to the purity of the enzyme in the specifications, in order to avoid cross-reactivity with other substrates. In some cases, impure enzymes can be used, such as jack bean urease and glucose oxidase for the food industry (General Mills), directly in a pseudo 'immobilised' form, e.g., as a liquid covered with a dialysis membrane. Consider also the activity of the enzyme − to be useful in an electrode, the activity should be at least 10 U/mg of protein.

Excellent tables of sources of enzyme can be found in references 4 and 8.

If the enzyme required for your electrode is not available commercially:

(i) Contact a large biochemical supply house and enquire whether it will isolate and purify the enzyme you want. Many will, for a suitable fee. New

England Enzyme Center of Tufts University (Boston, MA), for example, specialises in such a service.

(ii)   Look up the enzyme in the literature or standard biochemistry-enzymology reference books, ascertain the isolation and purification methods used, and perform the purification yourself. In most cases, the techniques specified are simple enough to be carried out by a person with reasonable scientific training.

Next, immobilise the enzyme, by one of the techniques described in Section 3.3, and assemble the electrode using the ion-selective electrode chosen (see also Section 3.3). Remember, the better the enzyme is immobilised, the more stable it is, and hence, the longer it can be useful and the more assays possible from one batch. The entrapment of soluble enzyme is simplest, and could be tried first. Physical entrapment or chemical attachment are the next methods of choice. The method recommended by the authors is glutaraldehyde attachment – it is simple, fast, and often gives highly desirable results.

3.3 Preparation of Electrodes

3.3.1 *Membrane-entrapped Enzymes*

Prepare the electrodes, as described by the diagram in *Figure 4*, using configuration A for physically entrapped enzymes and B for chemically bound or soluble entrapped enzymes.

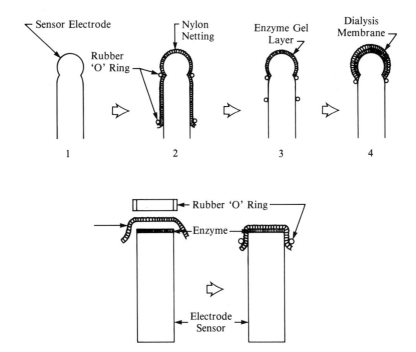

Figure 4. Preparation of enzyme electrodes. (**A**) With physically entrapped enzymes. (**B**) With chemically bound or soluble enzymes.

Table 2. Possible Electrode Sensors Useful in Construction of Enzyme Electrode.

	Useful for
A. *Potentiometric Sensors*	
NH_3	Urea, amino acids, glutamine, glutamic acid, nitrate, nitrite, creatinine, lyase and deaminase enzymes
CO_2	Urea, amino acids, decarboxylative enzyme systems
pH	Penicillin, RNA, DNA, glucose, enzyme reactions giving pH change.
I^-	Glucose, amino acids, cholesterol, alcohols
CN^-	Amygdalin
B. *Voltammetric (Amperometric) Sensors*	
O_2	Glucose, amino acids, organic acid, alcohols, uric acid, cholesterol, phosphate, all O_2 consuming enzymes
Pt or C	All redox enzymes, sulphate, uric acid, cholesterol, alcohols, glucose, amino acids, organic acids, NADH/NADPH systems

Choose the base sensor electrode, according to the enzyme reaction to be studied. It must respond to either one of the products or reactants of the enzyme system. For example, the base electrode and the types of enzyme electrodes that can be constructed from it are given in *Tables 1* and *2*. The methodology for the preparation of membrane-entrapped enzymes is described in *Table 3*.

3.3.2 *Chemically Attached Enzymes*

The enzyme is chemically attached to a solid support. This immobilised enzyme is placed onto the base sensor (according to *Figure 4B*), and covered with a dialysis membrane. Four methods which can be utilised are described. The glutaraldehyde and collagen membrane methods are the most easily effected, and are highly recommended (*Tables 4* and *5*). If denaturation of the enzyme occurs in immobilisation using either of these two methods (e.g., the enzyme electrode has very low activity), try the procedures described in *Tables 6* and *7*. The general chemical reactions for the preparations described in *Tables 4 − 7* are shown in *Figures 5 − 8*.

The procedure described for the preparation of collagen membrane electrodes (*Table 5*) is the same as reported by Coulet and co-workers (9), using a highly polymerised collagen, prepared under industrial conditions, as the binding site of various enzymes. Because of its protein nature, collagen has free amino groups available for covalent coupling with enzymes. The membranes also possess high polyol content (33%) which enhances film hydrophilicity, making it very supple and mechanically strong.

The preparation of covalently bound glucose oxidase *via* an acyl azide derivative of polyacrylamide or *via* a diazonium derivative of polyacrylic acid is shown in *Tables 6* and *7*, respectively.

Table 3. Preparation of Membrane-entrapped Enzymes.

Dialysis Membrane Electrode

1. Take the base electrode sensor, chosen from *Table 1* or *2*, and turn it upside down (see *Figure 4B*).

2. Place 10 – 15 units of the soluble enzyme, in the form of a thick paste of a freeze-dried powder, onto the surface of the base sensor (Note: for best results, the base sensor should be flat, not round).

3. Cover with a piece of dialysis membrane (cellophane of 20 – 25 μm thick, Will Scientific, Arthur H. Thomas, Sigma, etc.) about twice the diameter of the size of the electrode sensor.

4. Place a rubber O-ring, with a diameter that fits the electrode body snugly, around the cellophane membrane (*Figure 4B*), and gently push the O-ring onto the electrode body so that the enzyme forms a nice uniform layer on top of the electrode surface.

5. Place the electrode in buffer solution overnight to allow penetration of buffer into the enzyme layer and permit loss of entrapped air.

6. Store the electrode in buffer (optimum for the enzyme system) in a refrigerator between use.

Physically Entrapped Electrode

1. Place the electrode sensor upside down (*Figure 4A*) and cover with a thin nylon net (~ 50 μm thick – a sheer nylon stocking is satisfactory) which is secured with a rubber O-ring in the manner as shown. This serves as a support for the enzyme gel solution.

2. Dissolve N,N'-methylene-bis-acrylamide (Eastman Chemical Co.) (1.15 g) in phosphate buffer (0.1 M, pH 6.8, 40 ml) by heating to 60°C. Cool to 35°C, and add 6.06 g of acrylamide monomer (Eastman Chemical Co.), and filter into a 50 ml volumetric flask containing 5.5 mg of riboflavin (Eastman Chemical Co.) and 5.5 mg of potassium persulphate. This solution is stable for months if stored in the dark.

3. Prepare the enzyme gel solution by mixing 100 mg of enzyme (purity at least 10 U/mg) with 1.0 ml of gel solution. Gently pour the enzyme gel solution onto the nylon net in a thin film, making sure all of the pores of the net are saturated; 1 ml of this solution should be enough for many electrodes.

4. Place the electrode in a water-jacketed cell at 0.5°C, and remove oxygen, which inhibits the polymerisation, by purging with N_2 before and during the polymerisation. Complete the polymerisation by irradiating with a 150 W Westinghouse projector spot light for 1 h. The enzyme layer should be dry and hard.

5. Place a piece of dialysis membrane over the outside of the nylon net for further protection, and secure with a second O-ring.

6. Store the electrode in buffer in a refrigerator, overnight before use and between uses.

3.4 How to Use the Enzyme Electrode

Store the enzyme electrode in the buffer solution (type pH and optimum concentration dictated by the enzyme system used) in a refrigerator when not in use.

With soluble enzyme the electrode should be useful for about 50 assays over 7 days, with physically entrapped about 100 – 200 assays over 14 – 21 days, with chemically bound enzyme 200 – 1000 assays over 6 – 14 months. The dialysis membrane is added to:

(i) help hold the enzyme onto the electrode, and
(ii) keep bacteria out from the enzyme layer.

Table 4. Preparation of Glutaraldehyde-attached Enzymes.

1.	Turn the electrode base sensor upside down, as shown in *Figure 4B*.
2.	Place a piece of pig intestine membrane (Universal Sensors) onto the sensor, and secure with an O-ring.
3.	Add 50 μl of a solution containing phosphate buffer (0.1 M, pH 6.8, 2.7 ml), bovine serum albumin (Sigma Chemical Co.) (17.5%, 1.5 ml) and 50 mg of enzyme (at least 10 U/mg). Completely wet the membrane with the solution.
4.	Add 10 μl of a 10% solution of glutaraldehyde, rapidly mixing with a stirring rod. The enzyme layer should harden immediately.
5.	Let the membrane set overnight in a refrigerator, upside down, to ensure the layer is completely formed, then cover with a dialysis membrane, and store in the buffer solution used with the enzyme system.

Table 5. Preparation of Collagen Membrane Electrodes[a].

1.	Obtain collagen membranes from Centre Technique du Cuir, Lyon, France or Rutgers University, New Brunswick, NJ (diameter 2.5 cm and 0.1 mm thick in the dry state, 0.3 – 0.5 mm thick when swollen).
2.	Immerse these collagen membranes in 60 ml of 100% methanol containing 0.2 N HCl, for 3 days at room temperature.
3.	Take out five of these membranes, and rinse these thoroughly (the others can be kept in this solution) with distilled water.
4.	Place these into 100 ml of 1% hydrazine and keep immersed for 12 h at room temperature.
5.	Wash with distilled water at 0°C, then immerse into a mixture of 0.5 M KNO_2 and 0.3 N HCl (in an ice bath at 0°C) for 15 min.
6.	Wash the membranes with buffer solution (phosphate buffer, 0.05 M, pH 7.5), then place these membranes into a solution containing at least 50 Units of the enzyme in 2 ml of phosphate buffer (0.05 M, pH 7.5) and store in the refrigerator overnight at 4°C.
7.	Wash the membranes with buffer, and store in the same buffer until used.

[a]Upon use, the collagen enzyme layer is mounted onto the base electrode to be used (e.g., a CO_2 or NH_3 gas membrane electrode or a Pt electrode) and the enzyme layer is secured in place with a rubber O-ring of appropriate diameter. The electrode is placed into 2 ml of buffer containing the sample to be assayed (e.g., glucose) and the substrate concentration is measured from an appropriate calibration curve. After each analysis, the collagen membrane is washed several times with phosphate buffer, pH 7.5, 0.05 M.

To use the electrode, couple the enzyme electrode to a digital voltmeter, through the Universal Sensors Adaptor if using an amperometric probe (e.g., glucose) (see *Figure 3*). If the enzyme sensor does not have its own reference electrode built in plug a calomel electrode into the digital voltmeter.

Prepare two standard solutions of the substance to be assayed, at concentrations of 10^{-2} and 10^{-4} M in the buffer solution used. First place the enzyme and reference electrodes in the buffer solution to read the background potential, then insert the electrodes into the standard solutions and read the potential. Prepare a calibration plot of ϵ *versus* log concentration of substance to be assayed. Then insert the electrode pair into the solution(s) to be assayed, and read the potential. Calculate the concentration of the unknown solution(s) from the calibration

Table 6. Preparation of Covalently Bound Glucose Oxidase *via* an Acyl Azide Derivative of Polyacrylamide.

Preparation of Polyacrylamide Beads

1. Dissolve acrylamide (21 g), N,N'-methylenebisacrylamide (0.55 g), tris(hydroxymethyl)-aminomethane (Sigma Chemical Co.) (13.6 g), N,N,N',N'-tetramethylethylenediamine (Eastman Chemical Co.) (0.086 ml), and HCl (1 M, 18 ml) in 300 ml of distilled water.

2. Add ammonium persulphate as a catalyst. [Free radical polymerisation is brought about by stirring magnetically under a nitrogen atmosphere and irradiating with a 150 W projector spotlight (Westinghouse)].

3. After polymerisation is complete, remove the stirring bar and break the polyacrylamide into large pieces. Prepare small spherical beads of the polymer by blending at high speed with 300 ml water for 5 min. Store the resulting bead suspension under refrigeration.

Preparation of Hydrazide Derivative

1. Place 300 ml of the polyacrylamide bead suspension into a siliconised three-necked round-bottomed flask.

2. Immerse the flask and a beaker containing anhydrous hydrazine (Matheson, Coleman and Bell) (30 ml) in a 50°C constant-temperatue oil bath.

3. After about 45 min add the hydrazine to the gel. Stopper the flask, and stir the mixture magnetically for 12 h at 50°C.

4. Centrifuge the gel and discard the supernate. Wash the gel with 0.1 M NaCl by stirring magnetically for several minutes, centrifuging, and discarding the supernatant.

5. Repeat the washing procedure until the supernatant is esentially free of hydrazine as indicated by a pale violet color after 5 min when tested by mixing 5 ml with a few drops of a 3% solution of sodium 2,4,6-trinitrobenzenesulphonate (Eastman Chemical Co.) and 1 ml of saturated sodium tetraborate (Mallinckrodt). Store the hydrazide derivative under refrigeration.

Coupling of Glucose Oxidase

1. Place the hydrazide derivative (100 mg) in a plastic centrifuge tube and wash with 0.3 M HCl.

2. Suspend in HCl (0.3 M, 15 ml), cool to 0°C, and add sodium nitrite (1 M, 1 ml), also at 0°C.

3. After stirring magnetically in an ice bath for 2 min, rapidly wash the azide derivative formed with phosphate buffer (0.1 M, pH 6.8) at 0°C by centrifugation and decantation until the pH of the supernatant is close to 6.8.

4. Suspend the azide gel in phosphate buffer (0.1 M, pH 6.8, 10 ml) containing 100 mg glucose oxidase (Sigma Chemical Co., Type II, from *Aspergillus niger*).

5. Stir the mixture magnetically for 60 min at 0°C, after which time glycine (10 ml, 0.5 M) is added to couple with unreacted azide groups.

6. After stirring for an additional 60 min at 0°C, wash the enzyme gel several times with phosphate buffer (0.1 M, pH 6.8) and store under refrigeration.

curve. When the response of the enzyme electrode has deteriorated (see Section 4), remove the enzyme membrane, and replace it with a new membrane.

4. EXAMPLES OF ENZYME ELECTRODES

4.1 Commonly Used Enzyme Electrodes

Table 1 gives a listing of some enzyme electrodes that have been prepared for

Table 7. Preparation of Covalently Bound Glucose Oxidase *via* a Diazonium Derivative of Polyacrylic Acid.

Polymetrisation of Acrylic Acid

1. Dissolve approximately 50 ml of reagent-grade acrylic acid (Aldrich Chemical Co.) in 20 ml hexane and place in a round-bottomed flask.

2. Add a few milligrams of ammonium persulphate as a free radical initiator, and keep the system in a dry nitrogen atmosphere.

3. Heat the flask with a heating mantle until rapid polymerisation is observed. Quickly remove the mantle and allow the flask to cool to room temperature.

Preparation of Co-polymer

1. Break the polymer into small particles and neutralise with sodium hydroxide.

2. Evaporate the sodium salt to dryness in a rotary evaporator and grind to a fine powder.

3. Suspend the powder (~ 3.6 g) in 6 ml hexane and cool to approximately $4°C$;

4. Convert the acid to the acyl chloride by the addition of $SOCl_2$ (2.8 ml) with stirring in an ice bath for 1 h (removing generated gasses by suction).

5. Wash the acryl chloride polymer with ether and dry under vacuum.

6. Add nitroaniline (0.5 g) and ether (6 ml), and allow the mixture to stir overnight.

7. Filter the product formed, wash with ether, and air dry.

Coupling of Glucose Oxidase

1. Dissolve the *p*-nitroaniline derivative (150 mg) in 10 ml of distilled water and adjust the solution to pH 5 with dilute acetic acid.

2. Add ethylenediamine slowly with stirring until a fine white precipitate is observed.

3. Wash the precipitate three times with distilled water and suspend in 5 ml of distilled water.

4. Reduce the polymer by the addition of $TiCl_3$ and wash several times with distilled water by centrifugation and decantation.

5. Convert the reduced derivative to a diazonium salt by addition of nitrous acid (0.5 M, 10 ml) at approximately $4°C$ with stirring for 2 min.

6. Flush the diazonium salt intermediate with cold distilled water and rapidly wash several times with phosphate buffer (0.1 M, pH 6.8) by centrifugation and decantation.

7. Mix with a phosphate buffer solution (0.1 M, pH 6.8) containing 100 mg glucose oxidase (Sigma Chemical Co., Type II) at approximately $4°C$ for 1 h.

8. Wash the resulting gel several times with buffer and store under refrigeration.

---

analysis of common substrates together with the enzyme used, the sensor, the immobilisation method, the stability of the probe, the response time of the electrode, the units (U) of enzyme used to make the electrode and the range of concentrations determinable (see reference 4 for a complete listing of electrodes available).

In several cases, many different base sensors can be used. For example, for urea, one could use either a cation electrode, which measures the NH_4^+ ion formed in the urease-catalysed hydrolysis of urea:

$$\text{Urea} \xrightarrow{\text{Urease}} NH_4^+ + HCO_3^-$$

a pH sensor to measure the pH change due to the products formed, or a gas mem-

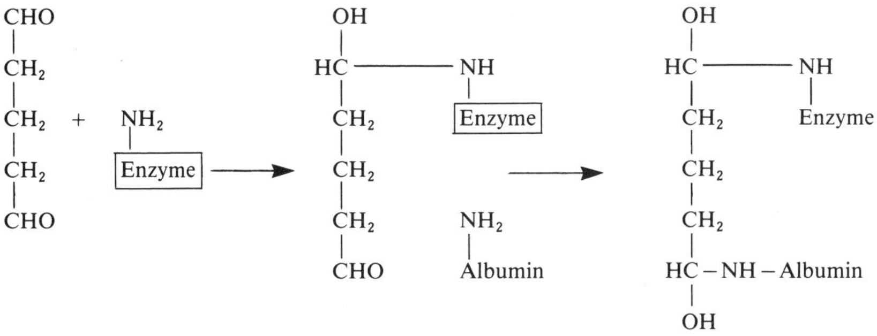

$$
\begin{array}{c}
\text{CHO} \\
| \\
\text{CH}_2 \\
| \\
\text{CH}_2 \quad + \quad \text{NH}_2 \\
| \qquad\qquad | \\
\text{CH}_2 \qquad \boxed{\text{Enzyme}} \\
| \\
\text{CHO}
\end{array}
\longrightarrow
\begin{array}{c}
\text{OH} \\
| \\
\text{HC}\!-\!-\!-\!-\!\text{NH} \\
| \\
\text{CH}_2 \qquad \boxed{\text{Enzyme}} \\
| \\
\text{CH}_2 \\
| \\
\text{CH}_2 \qquad \text{NH}_2 \\
| \\
\text{CHO} \qquad \text{Albumin}
\end{array}
\longrightarrow
\begin{array}{c}
\text{OH} \\
| \\
\text{HC}\!-\!-\!-\!-\!\text{NH} \\
| \\
\text{CH}_2 \qquad \text{Enzyme} \\
| \\
\text{CH}_2 \\
| \\
\text{CH}_2 \\
| \\
\text{HC}\!-\!\text{NH}\!-\!\text{Albumin} \\
| \\
\text{OH}
\end{array}
$$

Figure 5. Reactions involved in the preparation of glutaraldehyde-attached enzymes.

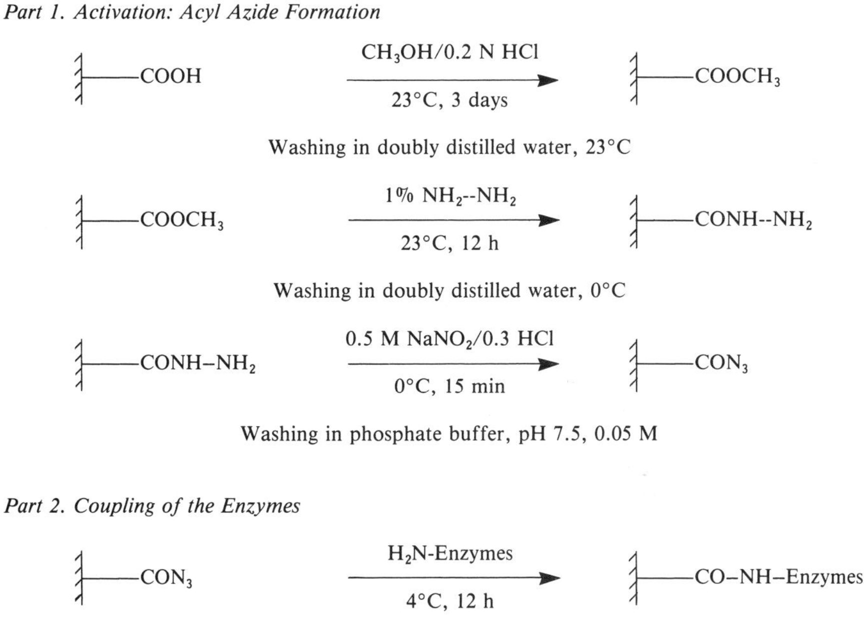

Part 1. Activation: Acyl Azide Formation

$$
\vdash\!\!-\!\text{COOH} \xrightarrow[\text{23°C, 3 days}]{\text{CH}_3\text{OH/0.2 N HCl}} \vdash\!\!-\!\text{COOCH}_3
$$

Washing in doubly distilled water, 23°C

$$
\vdash\!\!-\!\text{COOCH}_3 \xrightarrow[\text{23°C, 12 h}]{1\%\ \text{NH}_2\text{-}\text{-}\text{NH}_2} \vdash\!\!-\!\text{CONH-}\text{-}\text{NH}_2
$$

Washing in doubly distilled water, 0°C

$$
\vdash\!\!-\!\text{CONH}\!-\!\text{NH}_2 \xrightarrow[\text{0°C, 15 min}]{0.5\ \text{M NaNO}_2/0.3\ \text{HCl}} \vdash\!\!-\!\text{CON}_3
$$

Washing in phosphate buffer, pH 7.5, 0.05 M

Part 2. Coupling of the Enzymes

$$
\vdash\!\!-\!\text{CON}_3 \xrightarrow[\text{4°C, 12 h}]{\text{H}_2\text{N-Enzymes}} \vdash\!\!-\!\text{CO}\!-\!\text{NH}\!-\!\text{Enzymes}
$$

Figure 6. Reactions involved in the preparation of collagen membrane electrodes.

brane NH_3 or CO_2 electrode to measure either the NH_3 (formed by adding OH to $NH_4^+ \rightarrow NH_3$) or the CO_2 (formed by adding H^+ to the $HCO_3^- \rightarrow CO_2$). By far the best probe is the NH_3 electrode, because of its high specificity, and low limit of detection (10^{-6} M compared with 5 x 10^{-5} M for the CO_2 electrode). The disadvantage of using this electrode is slow response time ($2-4$ min) and long recovery time in return to the original base line ($5-10$ min). Guilbault and Mascini (10), for example, showed that by chemically attaching urease to a poly-propylene membrane, which is an integral part of the NH_3 gas membrane electrode, $200-1000$ assays can be performed on one electrode with a coefficient of variation (C.V.) of 2.5% over the range 5 x $10^{-5}-10^{-2}$ M. At least 20 assays/h can be made with excellent correlation with the results obtained by the spec-

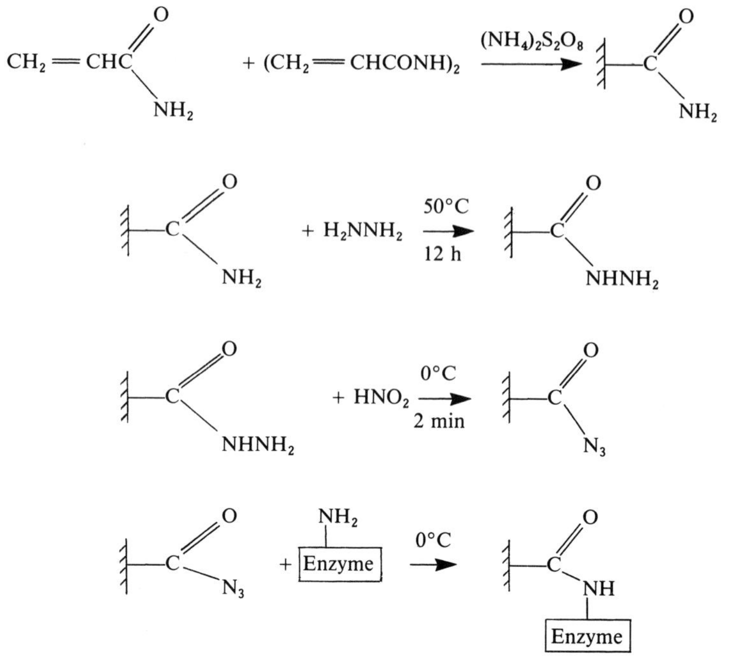

Figure 7. Reactions involved in the preparation of a covalently bound enzyme *via* an acyl azide derivative of polyacrylamide.

trophotometric diacetyl procedure.

The range of most enzyme electrodes is $10^{-2} - 10^{-4}$ M, with some extending up to 10^{-1} M (depending on the solubility of the substrate in the aqueous solution) and some extending down to 10^{-5} M, or lower, depending on the detection limit of the base sensor.

Deterioration of the enzyme electrode can be seen by three changes in the response characteristics:

(i)   with age the upper limit will decrease, from say 10^{-1} to 10^{-2} M;

(ii)   the slope of the calibration curve of potential *versus* log [concentration]; originally 60 mV/decade, Nernstian, will drop to 50, 40, perhaps 30 mV/decade, or lower, and

(iii)   the response time of the electrode, originally 30 sec to 4 min (approximately the same as that of the base sensor), will become longer as the enzyme ages.

In construction of an enzyme electrode, it is important that a highly purified enzyme be used (at least 10 U/mg) so that only a small amount of enzyme need to be used in the construction of the electrode. This will ensure a fast response time, approaching that of the base probe. As can be seen in *Table 1*, at least 10 U (1 mg) of enzyme is generally used.

The stability of the electrode depends on the type of entrapment, chemically attached enzyme electrodes being the most stable (500 – 1000 runs/electrode), with

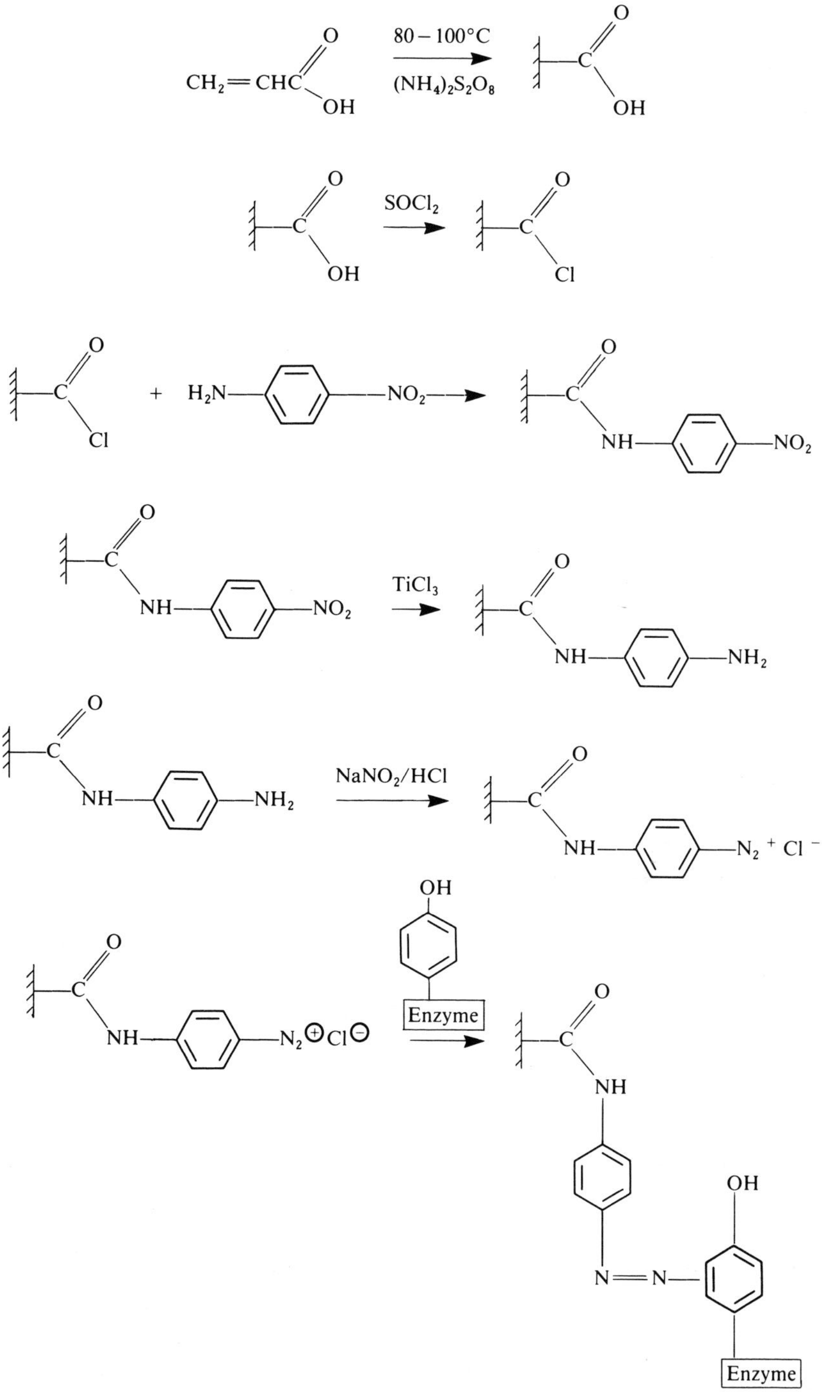

Figure 8. Reactions involved in the preparation of a covalently bound enzyme *via* a diazonium derivative of polyacrylic acid.

a storage stability of about 6 – 14 months.

The first enzyme electrode, an amperometric probe, was described by Clark and Lyons (1) who used a soluble glucose oxidase held between Cuprophane membranes. The oxygen uptake was measured with an O_2 electrode

$$\text{Glucose} + O_2 + H_2O \xrightarrow{\text{Glucose Oxidase}} H_2O_2 + \text{Gluconic acid}$$

Of all the base sensors that have been described to monitor this reaction (see *Table 1*: pH, Pt for H_2O_2 or quinone, I^- and gas O_2 electrodes), the best is still the O_2 electrode. The specificity is vastly improved by monitoring the initial O_2 uptake; most interferences in the assay of glucose arise from the reaction of the H_2O_2 formed with diverse substances, like ascorbate, vitamins, etc.

The first enzyme electrode for determination of L-amino acids were developed by Guilbault and Hrabankova (11), who placed an immobilised layer of L-amino acid oxidase over a monovalent cation electrode to detect the ammonium ion formed in the enzyme-catalysed oxidation of the amino acid. The general enzyme from snake venom, L-amino acid oxidase, permits the assay of L-cysteine, L-leucine, L-tyrosine, L-tryptophan, L-phenylalanine and L-methionine.

Two amino acids of special interest in food and nutrition are L-lysine and L-methionine. If the content of these two amino acids in the food can be accurately determined, the protein quality can be ascertained. This is of extreme importance in assessing the nutritive value of the food. However, to do this assay requires the separation of these two amino acids using amino acid analysers, and subsequent analysis using ninhydrin. Using the high specificity of enzymes, we have shown (see reference 4) that totally specific and sensitive electrode probes can be constructed for L-lysine using L-lysine decarboxylase from *Escherichia coli* on a CO_2 probe, and for L-methionine using L-methionine ammonia lyase onto an ammonia-specific electrode. From 10^{-2} to 10^{-5} M concentrations could be assayed with no interferences. Thus, no prior separation of amino acids is required.

Another very important electrode described has been the penicillin electrode, now widely used to monitor for the penicillin content of fermentation broths. The electrode is based on use of a pH probe, coated with immobilised enzyme penicillinase.

$$\text{Penicillin} \xrightarrow{\text{Penicillinase}} \text{Penicilloic acid}$$

The response time is very fast (< 30 sec) with a slope of 52 mV/decade over the range 5 x $10^{-2} - 10^{-4}$ M (reference 4).

4.2 Electrode Probes Using Whole Cells – Microbial Tissue Enzymes

One of the newest areas in biological electrode probes has been the application of whole cell microorganisms or tissue cells to the surface of an ion-selective electrode (I.S.E.) to form a bioselective sensor. Such a probe was first described by

Divies (12,13), and offers three main advantages.

(i)   Purified enzymes are not necessary — the whole cell or slice of tissue can be used directly without extensive purification and separation steps.

(ii) The electrode can be regenerated by immersion in nutrient broth. The microbe is essentially living and can be kept alive for long periods.

(iii) The whole cell can contain many enzymes and several cofactors that can catalyse extensive transformations that could be difficult, if not impossible, to effect with single immobilised enzymes. In addition, the cofactors necessary for enzymatic reaction are held in a natural immobilised state.

There are, however, some disadvantages:

(i)   poor selectivity can result because the bacteria or microbe contains several enzymes that can convert many different substrates, in addition to the one desired;

(ii) poor response times are often observed, because the enzymes in the microbe are present at low concentrations, and the electrode membrane is very thick subject to slow diffusion processes.

Some examples of analytical uses are *Neurospora europea* for ammonia, *Trichosporon brassicae* for acetic acid, whole living cells of *Sarcina flava* (glutaminase) for glutamine, *Azotobacter vinelandii* (nitrate reductase) for nitrate and sliced porcine kidney (glutaminase) for glutamine (see reference 4).

These electrodes can be prepared as described in Section 3.3, the preferred immobilisation method being physical entrapment (Section 3.3.1) or glutaraldehyde chemical coupling (Section 3.3.2).

4.3 Antigen-antibody Probes

Another possible application of biological probes is the construction of sensor probes utilising bound antigens or antibodies.

We have, for example, successfully immobilised creatine kinase M (CK-M) antibody as a pre-treatment for the detection of the cardio-specific CK-MB isoenzyme. Goat anti-human CK-M IgG was immobilised on a carrier (glass beads) through glutaraldehyde coupling, and the immobilised carrier was packed into a tube that could be used for several hundred assays, and is regenerable. Excellent results were obtained in an electrode developed for assay of CK-MB (see reference 4).

An alternative approach was presented by Suzsuki (14), who bound an antigen and developed an assay for syphillis, using its antibodies present in blood. The contact potential was measured, with very low mV changes (1 – 3 mV).

4.4 Commercial Availability of Probes

Immobilised enzymes, together with electrochemical sensors, are used in several instruments available commercially. Owens-Illinois (Kimble) has designed a urea instrument using immobilised urease and an ammonia electrode probe, and a glucose instrument using insolubilised glucose oxidase and a Pt electrode. Patent rights to this system have been purchased by Technicon, who markets the instrument in Europe.

Yellowsprings Instrument Co. (Ohio) markets a glucose instrument with an immobilised glucose oxidase pad placed on a Pt electrode, and has instruments available for triglycerides, lipase, cholesterol and amylase. Fuji Electric (Tokyo) has a glucose instrument, similar in design to the Yellowsprings Instrument.

Self-contained electrode probes are available from only two companies: Tacussel (Lyon France) which sells a glucose electrode based on the collagen immobilisation of Coulet, and Universal Sensors (PO Box 736, New Orleans, LA 70148), which offers probes for urea, glucose, creatinine, amino acids, alcohols, and others on special request, based on pig intestine immobilisation.

5. ACKNOWLEDGEMENTS

G.G. Guilbault was a Fullbright Scholar on leave from the Department of Chemistry, University of New Orleans, New Orleans, LA 70148, USA.

6. REFERENCES

1. Clark,L. and Lyons,C. (1962) *Ann. N.Y. Acad. Sci.,* **102**, 29.
2. Updike,S.J. and Hicks,G.P. (1971) *Nature,* **214**, 986.
3. Guilbault,G. and Montalvo,J. (1969) *J. Am. Chem. Soc.,* **91**, 2164.
4. Guilbault,G. (1984) *Handbook of Immobilized Enzymes,* published by Marcel Dekker, New York.
5. Guilbault,G.G. and Lubrano,G. (1973) *Anal. Chim. Acta,* **64**, 439.
6. Mascini,M. and Liberti,A. (1974) *Anal. Chim. Acta,* **68**, 117.
7. Anfalt,T., Granelli,A. and Jagner,D. (1973) *Anal. Lett.,* **6**, 969.
8. Guilbault,G.G. (1974) *Handbook of Enzymatic Analysis,* published by Marcel Dekker, New York.
9. Coulet,P.R., Thevenot,D.R., Strandberg,R., Laurent,J. and Gautharon,D. (1979) *Anal. Chem.,* **51**, 96.
10. Guilbault,G. and Mascini,M. (1977) *Anal. Chem.,* **49**, 795.
11. Guilbault,G. and Hrabankova,E. (1970) *Anal. Lett.,* **3**, 53.
12. Divies,C. (1975) *Ann. Microbiol. (Paris),* **126A**, 175.
13. Divies,C. (1976) *Chem. Eng. News,* **54**, 23.
14. Suzuki,S. (1979) *J. Solid Phase Biochem.,* **4**, 25.

CHAPTER 6

Electrochemical Techniques with Immobilised Biological Materials

LEMUEL B.WINGARD JR.

1. INTRODUCTION

Oxidation-reduction reactions involve the transfer of electrons between substrates and products and thus serve as one of the major types of reactions where the utilisation or transfer of significant quantities of energy is involved. Biological oxidation-reduction reactions as catalysed by oxidoreductase enzymes are especially intriguing because of the relatively mild reaction conditions and the lack of unexpected by-products, as compared with oxidation-reduction reactions carried out with non-enzymatic catalysts or without any catalyst. The heavy involvement of oxidation-reduction in industrial chemistry and also in the development of clinical chemical assay methods has led to strong interest in the utilisation of oxidoreductase enzymes in these two areas of application. In addition, basic research into the mechanisms of action of oxidoreductase enzymes continues as a major area of biochemical investigation. The conclusion from these introductory comments is that oxidation-reduction enzymes presently are under intense scrutiny both from the stand point of basic biochemical research as well as that of practical industrial and analytical chemistry applications.

Since electrons are transferred during oxidation-reduction reactions, electrochemical techniques can be used to characterise the type and extent of reaction and to help in identifying the products and suspected reaction pathways. In the case of oxidoreductase enzymes, the cofactor (C) portion of the holoenzyme actually undergoes oxidation or reduction and thus can be subjected to electrochemical characterisation along with the substrates (S) and products (P). This is shown schematically in Equation 1, where E_n' stands for the apoenzyme (holoenzyme E_n minus the cofactor) and C_{ox}, C_{re} represents the oxidised and reduced forms, respectively, of the cofactor.

$$S_1 + C_{ox} \xrightarrow{E_n'} P_1 + C_{rd}$$

$$C_{rd} + S_2 \xrightarrow{E_n'} P_2 + C_{ox}$$

$$\overline{S_1 + S_2 \xrightarrow{E_n'} P_1 + P_2}$$

Equation 1

In Equation 1, both the reactions to form P_1 and P_2 are shown as catalysed by E_n'; however, with some enzymes only the initial reaction to form P_1 is catalysed by E_n'. In practice the reaction to form P_2 and concomitantly to regenerate C_{ox} sometimes can be catalysed either by E_n' or by another enzyme and, in a few cases, by non-enzymatic materials. Electrochemically, one could monitor the holoenzyme, the cofactor, a substrate or a product. If the enzyme-catalysed reaction is being carried out in homogeneous solution, then all of the species present at least initially have equal access to the electrode surface. However, by immobilisation of one of the species on the electrode surface, several new possibilities arise. Since this book is concentrated on immobilised rather than free compounds or cells, the remainder of this chapter will be limited to methodology in which either the holoenzyme or the cofactor is immobilised on the electrode surface. However, comparisons with materials in solution will prove helpful in interpretation of the results.

The range of available electrochemical techniques for the study of immobilised oxidoreductase enzyme systems is extensive. The electrode containing the immobilised enzyme is placed in a solution of substrate plus enough electrolyte to give the solution a low electrical resistance. Then two principal options arise. The current can be controlled and the potential (chronopotentiometry) or quantity of charge (chronocoulometry) followed as a function of time. On the other hand, the potential can be controlled and the current (chronoamperometry) or its integral (chronocoulometry) followed *versus* time or the current examined simply as a function of the applied potential (voltammetry). In addition the electrode may be moving or stationary, and the solution may be quiet (non-stirred) or stirred. These techniques obviously cannot all be covered in a single chapter. Thus, I have picked two types of voltammetry, namely cyclic and differential pulse, to describe here since these methods are especially useful for characterising the mechanistic aspects of the reactions as well as for helping to identify the products for immobilised enzyme cofactor systems. The reader is referred elsewhere for the practical (1) or mathematical (2) aspects of other electrochemical methodology.

Immobilisation of the enzyme-cofactor system on the electrode surface, as compared with being free in solution, adds some additional variables that can be studied using voltammetry or other electrochemical methods. The possibility exists to control the spatial relationship of the apoenzyme and the cofactor relative to the electrode surface by using different coupling techniques, including alternative coupling positions on the enzyme or cofactor molecules and different types of linkages (3,4). Diffusional resistance for the transport of substrates or products through the immobilised enzyme-cofactor matrix also may be significant. For electroactive substrates or products, the magnitudes of the diffusional resistances can be measured using electrochemical techniques (5).

2. IMMOBILISATION OF OXIDATION-REDUCTION GROUPS ON ELECTRODE SURFACES

2.1. Description and Preparation of Supports

The support is the electrode surface and thus must be a reasonably good conductor of electricity. Platinum, glassy carbon, different forms of graphite, tin oxide, and a variety of pyridine-type polymers have been used widely in the preparation of chemically-modified electrodes (6) and thus are suitable supports for attachment of oxidation-reduction enzymes or cofactors. Metal electrodes, such as platinum, need to have an oxide film on the surface. Essentially any metal oxide will bind an alkyl silane. By proper choice of the functional group on the end of the alkyl chain, a variety of different linkages between the alkyl silane and the enzyme or cofactor can be utilised. For example, γ-aminopropyltriethoxysilane often is used for attachment of materials to platinum, tin oxide and other metal surfaces (6,7).

Glassy carbon is a gas-impermeable form of solid carbon that has a very low porosity and high strength. It is highly desirable for use as an electrode because it can be polished to give electrochemically reproducible surfaces (8). Most carbon surfaces contain low levels of various carbon-oxygen functional groups. With glassy carbon, the level of such groups normally is very low so that some type of surface activation is required. Oxidation of the glassy carbon surface to form carboxylic acids is usually the favoured method of activation; although in reality a variety of functional groups probably form (6). Functionalisation of glassy carbon can be carried out by electrochemical (9), radio frequency oxygen plasma (10) or hot acid (11) treatments with the level of functionalisation varying somewhat between research groups. The details are given in Section 2.2. We have obtained the highest loadings with the hot acid method of pre-treatment (12).

A variety of forms of graphite are used for electrode fabrication. Graphites in general are much more porous than glassy carbon and therefore are more difficult to utilise for preparing surfaces that are electrochemically identical. In addition, the high porosity gives a large pore surface area so that adsorption of organic compounds onto graphite surfaces is considerably stronger on graphite as compared with glassy carbon. Indeed, cofactors can be immobilised by adsorption on graphite electrodes (13).

The choice of electrode material is also influenced by the practical working range of applied potentials. For example, hydrogen evolution, electrolyte-electrode interactions and buffer-electrode interactions can produce enhanced background currents that limit the useful potential range for many types of electrodes. The practical range for platinum at highly acid pH is about -200 mV to $+1100$ mV; while at pH 7 the range is shifted to about -600 mV to $+600$ mV (14). For various forms of carbon the useful range is very roughly -1100 mV to $+1100$ mV (14).

2.2 Specific Immobilisation Techniques

This section contains a description of several detailed procedures for immobilisation of cofactors or enzymes on electrode surfaces. The intent is simply to provide a few examples since any attempt to cover all of the published methods would require many chapters.

Cofactors, such as flavine adenine dinucleotide (FAD) or nicotine adenine dinucleotide (NAD^+), or enzymes, such as glucose oxidase, can be immobilised by

adsorption on spectroscopic graphite as described in *Table 1*.
In the case of covalent attachment to glassy carbon (*Table 2*), Tokai Type
GC-A from International Minerals and Chemical Corp., in New York is a highly
desirable, but expensive grade. The material has a density of 1.50 g/cm³, and
1 − 2% porosity.

Table 1. Immobilisation of Cofactors or Enzymes on Graphite.

1. Polish the ends of the graphite rods[a] with 600 grit silicon cabide paper and then with a cloth impregnated with 1 μm diameter diamond powder.
2. Follow this by overnight Soxhlet extraction with ethanol or methanol to remove any adsorbed materials and then dry in a vacuum oven for several hours at room tempeature.
3. For adsorption, immerse the cleaned polished end of the graphite electrode for 12 − 24 h in a 1 mM aqueous solution of cofactor or a 1 − 20 mg/ml solution of the enzyme. Room temperature is used, unless the enzyme has low thermal stability, in which case 4°C is used.
4. Wash the loaded electrode with solutions of different ionic strength, pH and composition to dislodge loosely attached enzyme or cofactor without causing enzyme denaturation.

[a]Cylindrical pieces of Type 3751 SPK spectroscopic graphite, about 0.46 cm diameter by 3 cm long, can be obtained from the Carbon Products Division of Union Carbide Corp. The graphite has a density of 1.90 g/cm³, a porosity of 16.5%, and a mean pore diameter of 0.3 μm (13).

Table 2. Immobilisation of Cofactors or Enzymes on Glassy Carbon.

1. Cut the rods (with a diamond saw) into 2 − 3 cm lengths, polish the ends, clean and dry them as with the graphite in *Table 1*.
2. Oxidise the polished ends in one of the following ways:
   (a) treat for 10 min in a Harrick Scientific Corp. radiofrequency field at 150 milliTorr oxygen pressure;
   (b) treat for 15 min in 70% nitric acid followed by 60 min in 95% sulphuric acid at 170°C;
   (c) treat for 3 min in 10% nitric acid during which a potential of + 2.2 V with reference to Ag/AgCl (1 M KCl) is applied to the glassy carbon.
3. Wash the oxidised glassy carbon with water and activate the surface carboxylic acid groups using a 5 mg/ml solution of 1-ethyl-3-(3-dimethylaminopropyl) carbodiimide hydrochloride for 1.5 h at room temperature.
4. Wash the activated rods with water and incubate with the cofactor or enzyme solutions and subsequent washing to remove unattached material (12).

Table 3. Immobilisation of Cofactors or Enzymes on Platinum.

1. Soak the platinum electrode overnight in 15% nitric acid at room temperature for mild chemical oxidation; or apply a potential with the electrode in 0.5 N sulphuric acid and cycle the potential twice from + 1.90 V to − 0.45 V to + 1.90 V (ref. standard calomel electrode) at 50 mV/sec and then hold at + 1.90 V for 5 min for a heavier electrochemically-generated oxide film.
2. In either case, wash the oxide-coated electrode, dry it prior to attachment of an alkyl appendage.
3. Soak the oxide film for 1 h in a 10% solution of freshly distilled γ-aminopropyltriethoxysilane in anhydrous toluene (15).
4. Couple the silane amino group to an enzyme or cofactor amino group using, for example, 2.5% glutaraldehyde in 0.05 M sodium phosphate buffer pH 6.0 for 1 h at 0°C or another bifunctional reagent.

A typical procedure for activation of metal oxide surfaces involves cleaning the oxide films by overnight Soxhlet extraction with methanol followed by vacuum drying. An alternative cleaning method that also removes previously attached silanes is to place the platinum in boiling 1 N sodium hydroxide for 30 min followed by extensive rinsing with water and dilute nitric acid. The oxide film on the platinum surface to which cofactors and enzymes can be attached is generated chemically or electrochemically (*Table 3*).

Typical loadings of immobilised materials by the above methods are 10^{-9} -10^{-11} mol/cm^2. A loading of 10^{-10} mol/cm^2 is typical of monolayer coverage (6). The methodology for obtaining loading figures using voltammetry measurements is described in Section 3.3.

3. CYCLIC VOLTAMMETRY WITH IMMOBILISED MATERIALS

3.1 Description of Cyclic Voltammetry

In voltammetry, a potential is applied to the electrode, and the current is measured. The applied potential can be varied, for example, to give a linearly increasing value *versus* time. The resulting current-potential data becomes the equivalent of an absorbance-wavelength spectrum, with the amplitude, shape and relative positions of the current peaks important variables. In cyclic voltammetry, the applied potential is varied at a fixed rate in one direction and then switched in sign and varied at the same rate but in the opposite direction. A schematic diagram for cyclic voltammetry measurements is shown in *Figure 1* for the usual three-electrode system. Current flows from electrodes X-to-Y, with reference electrode Z present for the measurement of the applied potential. A typical plot of applied potential *versus* time is shown in *Figure 2*.

3.2 Apparatus and Procedures

A wide variety of cell designs are available in which to place the three electrodes. The cell needs to be jacketed for temperature control since the peak currents are temperature dependent. The cell also needs to be equipped for sparging with inert gases, such as nitrogen or argon, to exclude oxygen. The reference electrode is usually saturated calomel (s.c.e.), or silver-silver chloride (Ag/AgCl) containing KCl as the electrolyte. The reference (Z in *Figure 1*) and working (enzyme-cofactor) (X in *Figure 1*) electrodes must be placed very close together in order to

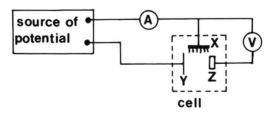

Figure 1. Schematic diagram for cyclic voltammetry, X is immobilised enzyme/cofactor electrode (called the working electrode). Y is an auxiliary electrode, and Z is a reference electrode against which the applied potentials are measured. A represents the measured current and V the applied potential.

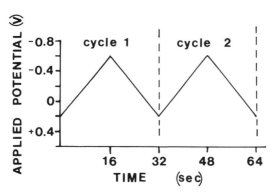

Figure 2. Typical applied potential *versus* time for cyclic voltammetry, shown for scanning between + 0.2 V and − 0.6 V at 50 mV/sec. The applied potential is with respect to a reference electrode.

minimise uncompensated solution resistances that might lead to errors in the applied potential. The auxiliary electrode (Y in *Figure 1*), typically a platinum wire or piece of platinum gauze, often is isolated from the other two electrodes by means of a glass frit to protect the working and reference electrodes from products produced at the auxiliary electrode. A potentiostat, such as the Princeton Applied Research Models 174A or 273 or the BAS Model 100, are suitable units for the generation of the applied potential for scan rates up to about 50 mV/sec and measurement of the resulting current. The current-potential plots are recorded on an X-Y plotter or at higher scan rates on a storage oscilloscope.

The procedure begins by mounting the electrodes in the electrochemical cell, adding electrolyte and pH buffers and substrates, and bubbling essentially oxygen-free nitrogen through the solution in the cell to remove dissolved oxygen. The cell interior is then blanketed with nitrogen to exclude oxygen. The potential scan is begun at a potential where none of the materials present will undergo oxidation or reduction. The direction of the initial potential scan will depend on the state of oxidation of the species present. If the material is in the oxidised form, then the first scan must be in the negative direction to generate a reduced species and give a reduction or cathodic current. On the other hand, starting with a reduced material will require an initial scan towards more positive potentials to give an oxidation or anodic current.

3.3 Results and Interpretation

It will be helpful firstly to discuss the results for the electroactive species in solution (*not* attached to the electrode surface) and secondly when attached to the electrode.

A typical cyclic voltammogram is shown in *Figure 3* for an electroactive species in unstirred solution. With an oxidised material present in solution, the scan would be started in this example at about + 0.2 V, and varied at 1 − 100 mV/sec in the direction of more negative potentials (i.e., stronger reducing capability). As the potential becomes more negative, it finally reaches a value (i.e., about − 0.20 V in *Figure 3*) where the oxidised cofactor, C_{ox}, begins to undergo reduc-

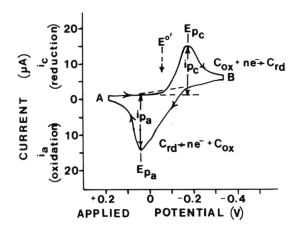

Figure 3. Typical current-potential plot of a single scan cyclic voltammogram for an electroactive species in solution taken between +0.20 V and −0.35 V. Arrows show the scan direction. Measurements started at A for an oxidised material or at B for a reduced material. Results shown are for a constant scan rate. See text for explanation of symbols.

tion. This transfer of electrons produces an increase in current. The current peaks at i_{pc} because the cofactor in solution in the vicinity of the electrode surface gets consumed (i.e., reduced), and because the rate of influx of additional oxidized cofactor from the bulk solution is diffusion controlled. At still more negative potentials, reduction current is limited by the rate of diffusion of oxidised cofactor to the electrode surface. The direction of the scan is reversed and the reduced cofactor now undergoes oxidation, in the *Figure 3* example at about −0.35 V. The formal potential $E^{o\prime}$ for the cofactor oxidation-reduction reaction is defined as follows:

$$C_{ox} + ne = C_{rd}$$

$$E = E^{o\prime} + \frac{RT}{nF} \ \frac{\ln[C_{ox}]}{[C_{rd}]}$$

Equation 2

where the items in brackets refer to concentrations. T is temperature, R the gas constant, F is the Faraday constant and E is the measured or actual potential. $E^{o\prime}$ is thus the measured potential (reference to a normal hydrogen electrode) with $[C_{ox}] = [C_{rd}]$. For cyclic voltammetry with the electroactive material in solution, as in *Figure 3*, the oxidation and reduction peak potentials, E_{pa} and E_{pc}, are separated by $0.059/n$ because the process is operating under diffusion controlled conditions. Thus, $E^{o\prime}$ is located midway between E_{pc} and E_{pa}. Here, n is the number of electrons transferred. The peak current, i_{pa} or i_{pc}, varies according to Equation 3 at 25°C for a reversible reaction.

$$i_{pc} = (2.69 \times 10^5) \, n^{3/2} \, A \, D_0^{1/2} \, v^{1/2} \, [C_0^*]$$

Equation 3

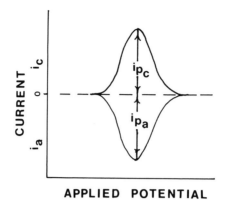

APPLIED POTENTIAL

Figure 4. Cyclic voltammogram for electroactive material attached to an electrode surface with none in solution and rapid, reversible electron transfer.

Where A is electrode area (cm^2), D_o is diffusion coefficient of oxidised cofactor (cm^2/sec), v is scan rate (V/sec), and $[C_o^*]$ is the bulk phase concentration of oxidised cofactor. Two points are especially relevant. First, i_p is proportional to the square root of the scan rate; and second, $E_{pa} - E_{pc} > 0.059/n$ if the process is not fully reversible.

When the electroactive species is attached to the electrode surface with no material free in solution, the cyclic voltammograms have a different shape. Since all of the oxidation-reduction material is on the electrode surface, we no longer have diffusional effects. Thus, the oxidation and reduction peaks are symmetrical and centred about $E^{o'}$, as shown in *Figure 4*. Another difference is that i_p is proportional to the scan rate to the first power for the attached case, as compared with the $1/2$ power for the solution case. Therefore the log-log plot of i_p *versus* scan rate is a good way of showing whether an electroactive species is in solution or immobilised on the electrode surface. The method of immobilisation can be covalent or ionic bonding, complexation or adsorption. For example, *Figure 5* shows a slope of 0.9 at low scan rates for FAD covalently attached to a glassy carbon electrode, as expected from the above discussion.

The above description of cyclic voltammetry results is based on the assumption that the rates of electron transfer are rapid compared with the scan rate of the applied potential. At higher scan rates this assumption is not valid, and kinetic rather than thermodynamically reversible conditions become controlling. The non-reversible conditions lead to increased separation between E_{pa} and E_{pc}, for both the solution and immobilised cases. The shifts in peak position with change in scan rate can be used to determine the kinetic rate constant; however, a discussion of that methodology is beyond the scope of this chapter. Such a shift is shown in *Figure 6A* for the cofactor FAD adsorbed on a glassy carbon electrode and measured in 0.1 M Tris buffer, pH 8.0, at 25°C and 50 mV/sec sweep rate.

In any electrochemical procedure where the applied potential is continuously undergoing change, a current is generated due to rearrangement of the electrical double layer adjacent to the electrode surface. This is called the charging current.

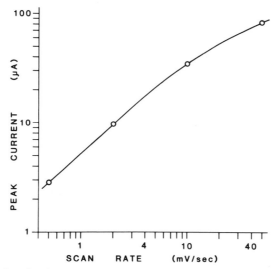

Figure 5. Log-log plot of peak current *versus* scan rate for FAD covalently attached to glassy carbon with 6-aminocaproic acid as a spacer. Reprinted with permission from Miyawaki and Wingard (12).

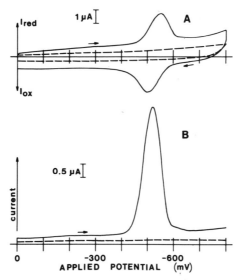

Figure 6. Cyclic (**A**) and differential pulse (**B**) voltammograms for FAD adsorbed on a glassy carbon electrode (——) and measured at 50 mV/sec in 0.1 M Tris buffer, pH 8.0, at 25°C under anaerobic conditions. Potentials with reference to Ag/AgCl. Control (----) results for bare glassy carbon electrode. Reprinted with permission from Miyawaki and Wingard (13). Copyright 1984, John Wiley and Sons Inc.

The component of i_p known as the faradaic current is the portion of the total current that is of interest; it must be measured from a baseline so chosen to exclude the charging current component of the total current. Since the double layer rearrangement is independent of diffusional effects, the magnitude of the charging current thus varies as the first power of the scan rate. The charging or baseline component of the measured current can become excessively large at high scan rates and with large surface area electrodes.

When the electroactive material is immobilised on the electrode surface, the total quantity of material can be assessed by integration over time of the faradaic component of the current peak. The total coulombs of energy transferred during the oxidation or reduction is obtained from the integral of current (ampere = coulomb/sec) with time and noting that 96 485 coulombs are transferred for each equivalent that is oxidised or reduced. From the equivalent weight, the moles or grams of the attached compound can be calculated. This quantity of immobilised material, when divided by the electrode area, gives the loading. Typical mono-layer coverage results in a loading of roughly 10^{-10} mol/cm^2. The area also can be determined electrochemically, using potassium ferrocyanide and normal pulse voltammetry. In this method very short duration pulses (a few milliseconds) are applied to the electrode about every 0.5 sec; and the resulting current is sampled. By linearly increasing the amplitude of the pulses at about 10 mV/sec, the potential for ferrocyanide oxidation is reached. This faradaic current can be used to calculate the electrochemically effective surface area of the electrode, which often is much larger than the geometrical area. By keeping the current pulses of short duration, only the ferrocyanide within a few atomic layers of the electrode surface is oxidised, thus increasing the accuracy of the method.

In summary, cyclic voltammetry can be used to show that the oxidation-reduction enzyme or cofactor is actually immobilised on the electrode surface by a look at log i_p *versus* log scan rate. E_p, i_p, the separation between E_{pa} and E_{pc}, and the peak shape as a function of scan rate are other important parameters. The most difficult task in the interpretation of cyclic voltammograms often is deciding where the baseline current lies from which to determine the value of i_p. However, the method has wide applicability and finds extensive usage.

4. DIFFERENTIAL PULSE VOLTAMMETRY WITH IMMOBILISED MATERIALS

4.1 Method and Apparatus

The sensitivity of cyclic voltammetry measurements can be increased markedly by using the derivative mode and by sampling the current at times when the charging current has decayed to zero. This is the essence of differential pulse voltammetry. The linearly increasing potential applied to the working electrode has super-imposed upon it a series of short duration pulses, as shown in *Figure 7A*. The same three-electrode system described in *Figure 1* for cyclic voltammetry can be used with the differential pulse mode, but the potentiostat must be equipped:

(i) for pulse generation;

(ii) for making current measurements just before the start of each pulse and again close to the end of each pulse; and

(iii) for giving as an output the difference in currents at each pulse.

The charging current, which is a maximum at the beginning of each pulse but decays to zero in a few milliseconds, is eliminated from the measurements by this selection of sample times.

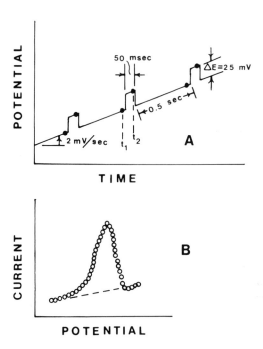

Figure 7. Differential pulse voltammetry. (**A**) applied potential as a function of time, with typical values shown for the scan rate and pulsing characteristics. (●) indicates when the current is sampled. For each pulse, the difference in current measured at t_1 and t_2 is plotted as a function of the applied potential, as shown by the (0) in (**B**). (---) in **B** is control current.

4.2 Results and Interpretation

A typical differential pulse voltammogram is shown in *Figure 7B*. With the electroactive material immobilised on the electrode surface, all of the material gets oxidised or reduced during the scan in potential, so that the current peaks and then drops nearly to zero. This is shown in *Figure 6B* for FAD adsorbed on glassy carbon. The peak is centred about the formal potential, $E^{o\prime}$; although kinetic factors may cause a shift in the location of the peak potential (16). With proper calibration, the peak height can be used to indicate the quantity of electroactive material immobilised on the electrode surface. The peak height also can vary with the magnitude of the uncompensated resistance of the solution or of the current measurement circuit, so that the calibration needs to be carried out under the same conditions as for the measurements (16).

In summary, differential pulse voltammetry is one of the most sensitive methods available for the detection of trace quantities of materials either in solution or immobilised on the electrode surface. For immobilised materials it may be difficult to calibrate the peak height so as to obtain quantitative information; however, the method is widely used to identify attached materials through the position of the peak potential.

5. DETERMINATION OF MASS TRANSFER RESISTANCES USING ROTATING ELECTRODES

Several electrochemical techniques for analysing electroactive species in solution are based on the slowest step being the rate of diffusion of electroactive material across the unstirred diffusion layer adjacent to the electrode surface. This was mentioned in the discussion of the cyclic voltammetry peak current values (see Equation 3, Section 3.3). Another technique which is based on diffusion limited current involves applying a linearly increasing potential to a rotating disc electrode. This technique is discussed further since it can serve as the basis for the measurement of diffusional resistances for the transport of an electroactive species through an immobilised enzyme matrix.

The rotating ring-disc electrode (RRDE) and associated circuitry is shown schematically in *Figure 8*. A Pine Instrument Co. potentiostat rotator, and removable platinum disc electrode makes an excellent unit for this type of study. The ring and disc are separated by Teflon insulation and are connected *via* lead wires to the potentiostat. A similar electrochemical cell with reference and auxiliary electrodes, as described earlier, is employed for carrying out voltammetry or other electrochemical techniques on either the ring or the disc. When the RRDE is spinning, centrifugal force conveys the product of any reactions occurring at the disc out to the ring where electrochemical detection can be employed. With an enzyme immobilised on the disc, the reaction products can be detected at the ring. In addition, the rate of diffusion of electroactive materials through the enzyme matrix attached to the disc can be assessed. For the covalent attachment of enzymes and cofactors to the platinum disc, non-removable discs can be employed when the enzyme/cofactor immobilisation temperature does not exceed 25°C ± about 10°C or when highly reactive reagents are not used. However, under

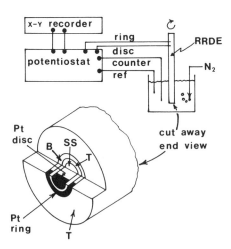

Figure 8. Description of rotating ring-disc electrode (RRDE). T is Teflon, SS is stainless steel, and B is brass. Disc diameter is 0.50 cm. Ring diameter is 0.75 cm inside and 0.85 cm outside. Reprinted with permission from Castner and Wingard (5). Copyright 1984, American Chemical Society.

greater temperature or chemical reactivity extremes, thermal contraction/expansion or chemical attack on the electrode may cause a leak at the disc-Teflon joint, thus leading to artifacts in the electrochemical response.

Covalent attachment of enzymes or cofactors to the platinum disc can be carried out by alkylamine silane activation of the platinum followed by glutaraldehyde cross-linking of the anchored amine with a similar group on the material to be immobilised, or allylamine can be used to activate the platinum surface (5). For alkylamine silane attachment of an enzyme, the platinum disc is cleaned, oxidised, removed and activated, as described in Section 2.2, by attachment of the aminosilane followed by linkage of the amino group to glutaraldehyde. Enzyme coupling can be accomplished by placing the activated disc in 2.5 mg/ml glucose oxidase at 4°C in sodium phosphate buffer 50 mM, pH 7.0 for 4 h. For attachment by the allylamine method, the disc is electrochemically pre-treated by scanning from $+400$ mV to $+1300$ mV at 0.5 V/min, then stepping the potential to -300 mV, and finally stepping to $+400$ mV and holding for 2 min. Immersion of the pre-treated disc in 4% aqueous allylamine allows the allyl group (double bond) to attach to the platinum oxide surface. Enzyme attachment is carried out by placing the disc in 1% glucose oxidase with the above buffer and about 1.25% glutaraldehyde for 4 h at 4°C. In either case the enzyme-loaded disc is washed with buffer, equilibrated to room temperature and reassembled in the RRDE.

With the enzyme, glucose oxidase attached to the platinum disc, the diffusion coefficient for transport of an electroactive material such as potassium ferrocyanide through the attached matrix can be measured electrochemically. In this case, the ring electrode is open circuited and not used. As the RRDE is rotated, the viscous drag at the disc surface causes the fluid near the metal surface to rotate. The resulting centrifugal force sweeps the fluid across the face of the disc in the radial direction. For a constant rotation speed, the rate of diffusion of electroactive material across the flow streamlines from the bulk solution to the electrode surface becomes the rate-limiting condition. Under these conditions the current at the disc is called the limiting current, i_l. It has been rigorously defined according to Equation 4 (ref. 2).

$$i_l = 0.62\, n\, \text{F}\, \text{A}\, \text{D}^{2/3}\, w^{1/2}\, \nu^{-1/6}\, [\text{G}_{rd}^*] \qquad \text{Equation 4}$$

where w is the rotation speed (rad/sec), ν is the kinematic viscosity (cm²/sec), and $[\text{G}_{rd}^*]$ is the bulk concentration of potassium ferrocyanide, F is the Faraday constant, A the area and D the diffusion coefficient. The other symbols have been defined earlier. The diffusion coefficient can be obtained from a plot of the limiting current *versus* the square root of the rotation rate at constant $[\text{G}_{rd}^*]$. The slope of the plot and Equation 4 are used to obtain the diffusion coefficient D.

With 1.95 M potassium ferrocyanide in 1 M KCl electrolyte solution at 25°C and the electrode rotating at a constant value, the potential applied to the disc is scanned linearly from -0.10 V to $+0.70$ V (ref. Ag/AgCl). The $E^{0\,\prime}$ for the ferrocyanide/ferricyanide oxidation-reduction couple occurs at about $+0.36$ V (ref.

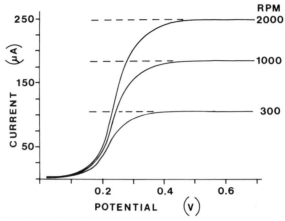

Figure 9. Oxidation of potassium ferrocyanide at rotating platinum disc electrode on which silane-glutaraldehyde-glucose oxidase is covalently immobilised. Bulk solution contains 1.95 M potassium ferrocyanide and 1 M KCl. Potential scanned at 2 V/min (ref. Ag/AgCl). Dashed lines show values of limiting currents. Reprinted with permission from Castner and Wingard (5). Copyright 1984, American Chemical Society.

normal hydrogen electrode) or about $+0.14$ V (ref. Ag/AgCl). Therefore, the ferrocyanide would undergo strong oxidation when the applied potential exceeded about 0.2 V (ref. Ag/AgCl). The results are shown in *Figure 9* for three different electrode rotation speeds. As the rotation speed is increased, the diffusion layer adjacent to the electrode surface becomes thinner thus allowing the rate of diffusion across this layer to increase. The concentration of ferrocyanide at the electrode surface is maintained at essentially zero, so that the concentration gradient and thus the rate of diffusion varies directly with the diffusion layer thickness. When the limiting current values from *Figure 9* are plotted along with data at other rotation rates, the fitted straight line follows the relationship of Equation 5, with a correlation coefficient of 0.999.

$$i_l = 15.9 \, w^{1/2} + 17.2 \qquad \text{Equation 5}$$

From the disc area of 0.196 cm^2 and $\gamma = 0.01$ cm^2/sec for water, the diffusion coefficient is found to be 0.57 ± 0.05 x 10^{-5} cm^2/sec. This is slightly less, as expected, than for potassium ferrocyanide diffusion measured in water with a clean platinum disc (0.61 ± 0.03 x 10^{-5} cm^2/sec) (5).

In addition to using electrochemical techniques to determine the diffusion coefficient for mass transfer of an electroactive material through an immobilised enzyme matrix, electrochemical techniques also can be employed for determining the kinetic constants of immobilised enzymes. The entire method is too complex to explain here in detail (5). Basically the substrate for the enzymatic reaction is placed in the solution in which the RRDE is rotated. The enzyme, for example glucose oxidase, is immobilised on the disc and the RRDE rotated at constant speed. The key necessity is for one of the products to be electroactive. In the case of glucose oxidase, hydrogen peroxide is a product and is electroactive. By stepping the potential of the ring from a value (-0.02 V) where H_2O_2 is not affected to a value ($+0.80$ V) well in excess of the potential needed to produce rapid oxidation, the rate of generation of H_2O_2, i.e., the rate at which H_2O_2 reaches the ring, can be determined from the ring current. In a manner similar to the Eadie Hofstee method for linearising the single substrate Michaelis-Menton equation for enzyme kinetics in terms of reaction velocities, the equation can be presented in terms of current (Equation 6).

$$i_{ss} = i_{max} - K'_m \left(\frac{i_{ss}}{[G_g^*]} \right) \qquad \text{Equation 6}$$

Where i_{ss} is the steady-state current measured a few minutes following the step in potential applied to the ring, i_{max} is the steady state current when the glucose oxidase is operating at maximum velocity, K'_m is the apparent Michaelis constant, and $[G_g^*]$ is the bulk solution concentration of glucose. Plots of i_{ss} *versus* $i_{ss}/[G_g^*]$ have been shown to be linear for immobilised glucose oxidase (5), thus demonstrating the usefulness of the method.

6. SUMMARY AND CONCLUSIONS

Cyclic and differential pulse voltammetry are widely used qualitative and, to some extent, quantitative electrochemical methods for the characterisation of electroactive species attached to the surface of solid electrodes. The differential mode in particular has very high sensitivity such that less than monolayer coverage of the electrode surface can be examined. Additional electrochemical methods, of which only two are mentioned in this chapter, can be used to measure the mass transfer resistance imparted by the immobilisation matrix. As a wide variety of enzyme and enzyme-cofactor electrodes are developed, electrochemical methodology undoubtedly will find wider use. The methods are essentially non-destructive to the electrodes so that the same immobilised enzyme or cofactor electrode can be tested electrochemically and then utilised in other types of studies.

7. ACKNOWLEDGEMENTS

This work was supported by grants from the National Institute of Health (R01AM26370), National Science Foundation (CPE8107715), and Army Research Office (DAAG2982K0064). The efforts of Dr. James F. Castner, Dr. Osato Miyawaki, and Dr. Krishna Narasimhan in carrying out the studies are greatly appreciated.

8. REFERENCES

1. Kissinger,P.T. and Heineman,W.R., eds. (1984) *Laboratory Techniques in Electroanalytical Chemistry,* published by Marcel Dekker, New York.
2. Bard,A.J. and Faulkner,L.R. (1980) *Electrochemical Methods: Fundamentals and Applications,* published by John Wiley, New York.

3. Wingard,L.B.,Jr., (1980) in *Enzyme Engineering: Future Directions,* Wingard,L.B.,Jr., Berezin,I.V. and Klyosov,A.A. (eds.), Plenum Press, New York, p. 339.
4. Wingard,L.B.,Jr., (1983) in *Proc. Biotech 83,* Online Pubications Ltd., Northwood (UK) p. 613.
5. Castner,J.F. and Wingard,L.B.,Jr., (1984) *Biochemistry (Wash.),* **23,** 2203.
6. Murray,R.W. (1980) *Acc. Chem. Res.,* **13,** 135.
7. Wingard,L.B.,Jr., Ellis,D., Yao,S.J., Schiller,J.G., Liu,C.C., Wolfson,S.K.,Jr., and Drash, A.L. (1979) *J. Solid Phase Biochem.,* **4,** 253.
8. Van der Linden,W.E. and Dieker,J.W. (1980) *Anal. Chim. Acta,* **119,** 1.
9. Engstrom,R.C. (1982) *Anal. Chem.,* **54,** 2310.
10. Miller,C.W., Karweik,D.H. and Kuwana,T. (1981) *Anal. Chem.,* **53,** 2319.
11. Wingard,L.B.,Jr., and Gurecka,J.L.,Jr., (1980) *J. Mol. Catal.,* **9,** 209.
12. Miyawaki,O. and Wingard,L.B.,Jr., (1985) *Biochim. Biophys. Acta,* **838,** 60.
13. Miyawaki,O. and Wingard,L.B.,Jr., (1984) *Biotechnol. Bioeng.,* **26,** 1364.
14. Adams,R.N. (1969) *Electrochemistry at Solid Electrodes,* published by Marcel Dekker, New York.
15. Narasimhan,K. and Wingard,L.B.,Jr., (1985) *Enzyme Microbial Technol.,* in press.
16. Brown,A.P. and Anson,F.C. (1977) *Anal. Chem.,* **49,** 1589.

CHAPTER 7

Immobilised Cells:
Transformation of Steroids

K.A.KOSHCHEYENKO and G.V.SUKHODOLSKAYA

1. INTRODUCTION

During the last decade there has been increased interest in the field of microbial transformation and biosynthesis of organic compounds on the strength of new methods, specifically immobilisation of microbial cells. The use of immobilised microbial cells (IC) as biocatalysts has a number of advantages over both free cells and immobilised enzymes (1 − 4). A high efficiency of IC provides a wide range of successful applications in one-stage and multi-enzyme processes for production of amino acids, organic acids, antibiotics, enzymes, coenzymes and carbohydrates (1,3). IC are at present being investigated widely regarding their application in the purification of industrial wastes from organic and inorganic compounds (1,3). In recent years, IC have been successfully employed for analytical purposes (3,5).

Much attention is paid to the use of IC for different modifications of steroid compounds. It is well known that some steroid-transforming enzymes, e.g., dehydrogenases and hydroxylases, are highly labile proteins, their isolation and purification is laborious work. Immobilised cells are active multi-enzymic systems regenerating cofactors.

Such advantages make IC highly attractive for steroid transformation, and steroid compounds have become the first substrates transformed by IC. The first paper on this subject was published in 1970; it was devoted to 11β-hydroxylation of cortexolone by *Curvularia lunata* entrapped in polyacrylamide (6).

The analysis of papers dealing with different modifications of steroid compounds by IC shows that most investigators focused their attention on commercially important processes: Δ^1-dehydrogenation, 11β- and 11α-hydroxylation and transformation of sterines to C_{19}- and C_{22}-steroids (1 − 4, 6 − 10). One can distinguish several trends of investigations devoted to the creation of highly active and stable immobilised systems for steroid transformation, including the use of damaged and intact microbial cells. A certain amount of success has been achieved in the search for natural and synthetic carriers used to immobilise bacterial, yeast and fungal cells as well as to carry out processes in aqueous-organic two-phase systems (11 − 14).

The study of morphological and physiological-biochemical peculiarities of IC during immobilisation and transformation shows good prospects for the use of

91

living IC.

The natural localisation and spatial interrelationship of enzyme systems in an intact cell ensures their stability. In this respect the maintenance of cell viability during immobilisation and subsequent use of the carrier is a matter of topical interest.

This chapter presents the experimental data on the use of living cells of *Arthrobacter globiformis* 193 to carry out $\triangle^1$-dehydrogenation, $\triangle^1$- and 20β-reduction, *Saccharomyces cerevisiae* VKM y-483 to carry out 17β-reduction, and *Tieghemella orchidis* IBPhM-223 to carry out 11β- and 11α-hydroxylation of steroid compounds, and considers possible means of maintaining IC in the viable state.

2. MATERIALS AND METHODS USED IN STEROID TRANSFORMATION

The following reagents were used: hydrocortisone (4-pregnene-11β, 17α, 21-triol-3,20-dione) from 'Rusel Uclaf' (France) and 'Akrikhin' (USSR), prednisolone (1,4-pregnadien-11β, 17α, 21-triol-3,20-dione), cortisone (4-pregnene-17,21-diol-3,11,20-trione), prednisone (4-pregnadien-17α-21-diol-3,11,20-trione), cortisone acetate, prednisone acetate, cortexolone (Reichstein's substance 'S') 11-deoxycortisol; 4-pregnene-17α,21-diol-3,20-dione from 'Akrikhin' USSR), Sekosteroids: sekoketon-3-metoxy-$\triangle^{1,3,5,(10),9(11)}$ 8,14-sekoestratetraendiol-3,17β-on-14. Androstenedione(4-androstene-3,17-dione), androstadienedione(1,4-androstadiene-3,17-dione) were obtained from the Institute for Biorganic Chemistry (USSR, Academy of Sciences).

Acrylamide (AA), N,N'-methylenebisacrylamide (MBA), ammonium persulphate, N,N,N',N'-tetramethylethylenediamine (TEMED) were purchased from 'Serva' (FRG) and 'Reanal' (Hungary). Acrylamide from 'Reanal' was recrystallised once or twice before use. $\varkappa$-Carrageenan was purchased from 'Serva' and sodium alginate was from British Drug Houses.

Adsorbents for cells were powdered cellulose TU-6-09-3575-74, 0.5 mm fraction, cellulose for chromatography FND-9303, microcrystalline cellulose LK for chromatography ('Chemapol' Czechoslovakia), diethylaminoethylcellulose (DEAE), silicagel 'Chemapol L' (Czechoslovakia) 40/100 μm, polyvinyl alcohol (PVA-GOST-150214MPGU6-09-400467).

Photosensitive polymers: polyethyleneglycoldimethacrylate, the gift of Professor S.Fukui (Kyoto University, Japan); β-nicotinamide adenine dinucleotide reduced, phenazine methosulphate (PMS), menadione, Tween 20, 40, 60, 80, SPEN 40, 60, 80, SDS, cetyltrimethylammonium bromide, chloramphenicol ('Serva').

Glutaraldehyde, osmium-VIII oxide, cocadylsäure·Na-salz, uranyl acetate, lead citrate ('Serva'), Araldit M, Araldit Hardener and DMP-30 ('Fluka', Sweden).

For chromatography, preparation of media, and for other purposes 'Chemically pure' and 'Pure' reagents (USSR) were used.

2.1 $\triangle^1$-Dehydrogenation of Steroid Compounds

Among different modifications of steriod compounds, the reaction of microbial $\triangle^1$-dehydrogenation is the most important from a practical point of view, since the double bond introduction into steroids causes radical changes in their physiological activity.

We used the culture A. *globiformis* 193 for the Δ^1-dehydrogenation and 20-keto group reduction, Δ^1-hydrogenation, de-acetylation and complete destruction of a steroid molecule. The character of these enzymatic transformations is conditioned by the total combination of the factors: composition and redox potential of the medium, biomass, aeration and mass transfer conditions (15,16).

This part of the chapter presents the experimental methods used for the Δ^1-dehydrogenation of hydrocortisone and some other steroid compounds by IC (*Scheme 1*).

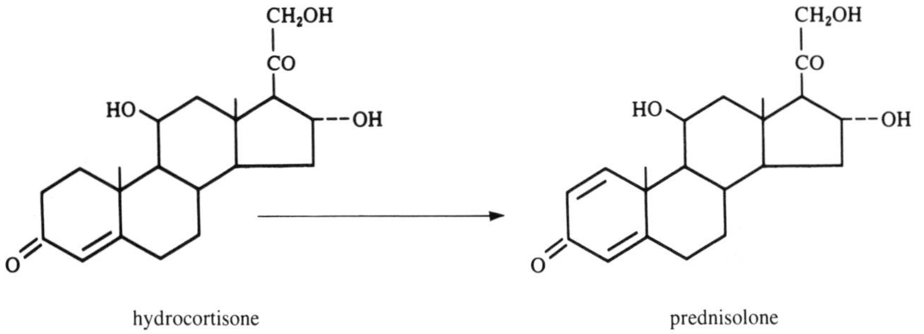

Scheme 1. Transformation of hydrocortisone to prednisolone.

The reaction of Δ^1-dehydrogenation is catalysed by the intracellular enzyme 3-ketosteroid-Δ^1-dehydrogenase (EC 1.3.99.4). In A. *globiformis* 193 this enzyme is localised in the cytoplasmic membrane, probably on its outside, i.e., the side facing the periplasm (17,18). Intact bacterial cells and the preparation of cytoplasmic membranes obtained by their ultrasonication carry out Δ^1-dehydrogenation of hydrocortisone to prednisolone which involves oxygen consumption. Δ^1-Dehydrogenation is inhibited by cyanide and 2-n-nonyl-4-hydroxyquinoline-N-oxide. Oxidation of hydrocortisone is accompanied by the reduction of cytochromes both in intact cells and preparations of cytoplasmic membranes. Under aerobic conditions, Δ^1-dehydrogenation of hydrocortisone generates energy in the form of a transmembrane potential. This points to the fact that electron transfer from 3-ketosteroid-Δ^1-dehydrogenase to oxygen occurs along the respiratory chain (17,18).

Therefore, Δ^1-dehydrogenation of hydrocortisone is a complex process which involves 3-ketosteroid-Δ^1-dehydrogenase and the electron-transport chain. The rate of hydrocortisone oxidation is controlled precisely at the stage of electron transfer along the respiratory chain. The process of transformation is not limited by the transfer of reagents through the cell wall and cytoplasmic membrane (17).

Transformation of hydrocortisone to prednisolone and oxygen uptake occur stoichiometrically (2 mol prednisolone per mol oxygen uptake). The kinetics of Δ^1-dehydrogenation of hydrocortisone obey the Michaelis-Menten equation. The apparent K_m value is 90 μM, V_{max} is 165 nmol/mg/cells/min with respect to hydrocortisone (17).

Table 1. Growth Conditions for the Culture *A. globiformis*.

Growth on solid nutrient medium		*Growth in liquid nutrient medium*				
		Inoculate		*Transforming culture*		
Composition of medium, %	*Duration, conditions*	*Composition of medium, % volume, ml*	*Duration, conditions*	*Composition of medium, % volume, ml*	*Duration, conditions*	
Corn-steep extract-1 glucose-1 agar-agar-3 pH 6.8−7.23	3−4 days at 29°C	Corn-steep extract-1 glucose-1 pH 6.8−7.2 150 ml	2−3 ml of cell suspension are introduced (D = 0.35−0.4 when diluted with water 1:9, 0.5 cm cuvette) 24 h growth at 220 r.p.m., *t* = 29°C.	Corn-steep extract-1 glucose-1 pH 6.8−7.2 150 ml	10 ml of inoculate are introduced. 24 h growth at **220 r.p.m.**, *t* = **29°C**. To induce 3-ketosteroid-△¹-dehydrogenase, 20 mg of acetate cortisone in 1 ml ethanol are introduced simultaneously per 100 ml medium	

The analysis of the experimental data which characterise the mechanism of $\triangle^1$-dehydrogenation performed by *A. globiformis* highlights the necessity for long-term maintenance of IC cells in the viable state in order to preserve their normal metabolic activity allowing continuous functioning of the 1,2-dehydrogenase system. The medium composition and culture conditions are given in *Table 1*. After a $22-24$ h period of growth separate the culture from medium by centrifugation at 5000 g for 15 min, and wash with 10 mM sodium phosphate buffer. Resuspend the bacteria to given a cell concentration of $17-100$ mg cells (dry weight) per 1 ml of buffer solution.

The bacterial suspension can be immobilised using the following techniques: entrapment in polyacrylamide, $\varkappa$-carrageenan, calcium alginate, agar, membranes of polyvinyl alcohol, photocross-linkable polymers and protein membranes; adsorption on various cellulose and large pore ceramics; binding on activated silica gel.

2.2 Entrapment in Polyacrylamide

Polyacrylamide, widely used to immobilise microbial cells, is a product of acrylamide co-polymerisation with a linking agent, usually MBA. Ammonium persulphate is used as an initiator and TEMED as a catalyst (19).

Polymerisation is carried out in a thin-walled beaker (bottom 10 cm in diameter) which is placed in a crystalliser filled with ice.

(i)   Introduce preliminary cooled solutions of reagents listed in *Table 2* and the cell suspension in succession into the beaker and gently stir. The beginning of gel formation is judged by the turbidity of the polymerisation solution and increase in the viscosity of the mixture which is observed $1-2$ min after the mixing of all the reagents.

(ii)   Place the beaker in a refrigerator for $10-15$ min at $-4\,^\circ$C for completion of the process of gel formation (the gel block formed should be 0.6 cm high). Even with the larger volume of polymerisation mixture, the above-mentioned gel height should not be exceeded in order to stabilise the temperature regime. It is not recommended that cells remain in gel for more than 2 h when kept at room temperature.

(iii) Mechanically fragment the gel block containing cells using a sieve (holes $0.8-1$ mm in diameter). Decant the suspension of the granules so obtain-

Table 2. Reagents Used to Prepare Polyacrylamide Gels.

Monomer solution	Cell suspension	Solution of a polymerisation inducer	Solution of a poly-merisation catalyst
Acrylamide 1.9 g MBA 0.1 g Distilled water 11 ml	6 ml of cell suspension in buffer solution	Ammonium persulphate 15 mg Distilled water 3 ml	TEMED (0.1 ml)

[a]All reagents are from 'Serva', AA — recrystallised.
Ammonium persulphate solution should be fresh.
The total volume of the polymerisation mixture is 20.1 ml.
Polymerisation can be performed both in air and in an inert gas atmosphere.

ed repeatedly with 4 − 5 litres of 0.01 M sodium phosphate buffer (pH 7.2) or physiological solution to select granules equal in size and to remove cells which fail to entrap.

In studies of morphophysiological peculiarities of IC, all the procedures including cultivation, preparation of bacterial suspension, immobilisation, preparation of granules, transformation and incubation of granules in the nutrient medium, should be carried out under strictly aseptic conditions.

The growth of biomass entrapped in the gel after incubation in the nutrient medium is determined by the gravimetric method, and by the analysis of protein (20). It has been shown that, after the entrapment of 100 mg and 600 mg cells (dry weight) to the gel block, 1 ml of gel granules contains 1.9 and 11 mg cells, respectively.

2.2.1 Steroid Transformation by Polyacrylamide Gel-entrapped Cells: the $\triangle^{1}$-Dehydrogenation Reaction

Steroid compounds are transformed by cells entrapped in gel granules in three types of reactors, as shown in *Figure 1*: packed-fed column reactor (type 1), continuous-flow stirred reactor (type 2) and stirred reactor (type 3).

In type 1 and 2 reactors, transformation of steroid compounds by IC occurs continuously, type 3 reactors work in a batch-wise manner: at the end of transformation the product is removed, the carrier with cells is washed repeatedly with

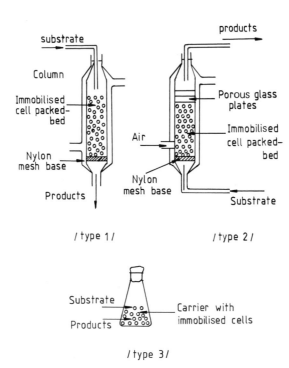

Figure 1. Reactors for transformation of steroid compounds.

buffer or physiological solutions and is re-used for transformation. Between transformations, granules can be kept in a refrigerator. Lyophilisation is advisable if a long storage (a month or more) of granules with active cells is required.

(i) Carry out transformation in 750 ml shaking flasks (180 − 220 r.p.m., $t=29°C$) with 9 − 18 ml granules (1 ml of granules contains 1.9 or 11 mg cells) and substrate (see below).

(ii) Add 10 mM phosphate buffer (pH 7.2) to adjust the liquid phase to 100 ml.

The substrate is hydrocortisone taken at a concentration of 0.2 mg/ml for type 1 reactor [ethanol concentration 2% (w/v)], and at 0.2 − 1 mg/ml for type 2 reactor [ethanol concentration 2 − 4% (w/v)] or in a microcrystalline state (particles 10 − 50 μm in size at a concentration of 0.5 − 50 mg/ml) for type 3 reactor.

Steroid compounds are analysed by thin-layer chromatography (t.l.c.). Samples (2 − 5 ml) are taken at intervals to determine the products formed during transformation. They are extracted three times with a double volume of chloroform, the solvent is evaporated under vacuum, and dry sediment is dissolved in ethanol to a concentration of 2 μg/μl. 20 − 40 μg samples are applied to silufol UV-254 plates. Steroids are separated in the solvent benzol:acetone (3:1). Before using, plates should be washed with ethanol for 20 − 30 min to remove substances absorbing the u.v. light.

To attain a clear-cut separation of steroids, a second chromatographic treatment is advisable. After separation, plates are air-dried and regions that absorb the u.v. light are marked. These are cut out together with a support and placed in test-tubes containing 5 ml ethanol. Elution is carried out for 15 min with periodic shaking; the optical density of the eluate is measured on a spectrophotometer at 240 nm.

The quantity of steroids can be determined by comparison with standard samples. To do this, a standard sample containing a known quantity of a certain steriod (1 − 40 μg) is applied alongside a sample. By visual comparison of the standard and the sample spot sizes and intensities, it is possible to quantify substances with an accuracy of 1 − 2% in the range 0.5 − 2 μg and 5 − 10% in the range 10 − 40 μg.

The enzymic activity of IC under continuous and batch-wise conditions of transformation is estimated from the quantity of the product obtained which is expressed as a percentage of the initial quantity of the substrate, or in mol/min.

2.2.2 *Effect of Immobilisation Conditions on $\triangle^1$-Dehydrogenase Activity*

When cells are entrapped in gel under the conditions described above, 3-ketosteroid- $\triangle$ ¹-dehydrogenase activity remains unaltered. In this case the specific $\triangle^1$-dehydrogenase activity of IC corresponds to that of free cells and makes up 0.14 μmol/mg cells/min. Changes in the immobilisation conditions such as increased temperature, time of gelling and increased polyacrylamide concentration exert a negative effect on the enzymic activity (21). For example, as seen in *Table 3*, cell contact with the polymerisation reagents for 20 min at 4°C (variant 2) results in the death of a significant quantity of cells. Only 6.4% of the cells remain viable under such conditions compared with variant 1. Elevation of the

Table 3. Effect of Immobilisation Conditions on the Viability of *A. globiformis* Cells.

Variant	Duration (min)	Temperature °C	Specific activity μmol/mg cells/min	Viability, %
1	1–2	4	0.14–0.15	100
2	20	4	0.038	6.4
3	20	26	0.009	0.5

temperature to 26°C, for the same duration of immobilisation, decreases still further the number of viable cells (variant 3) which is now only 0.5%. Therefore:

(i)   3-ketosteroid-$\triangle^1$-dehydrogenase activity clearly depends on the viability of IC;

(ii) 3-ketosteroid-$\triangle^1$-dehydrogenase activity is maximal provided the bulk IC remains viable due to a short polymerisation time (1 – 3 min) at 4°C;

(iii) viability of IC is determined by immobilisation conditions.

2.2.3 *Repeated* $\triangle^1$-*Dehydrogenation by Immobilised A. globiformis*

The ability of *A. globiformis* immobilised cells to transform hydrocortisone repeatedly is studied under the conditions described in Section 2.2.1 (type 3 reactor) using 9 ml of granules (17 mg of cells) at a concentration of hydrocortisone of 1 mg/ml.

Under such transformation conditions it is possible not only to preserve the activity of IC equal to that of free cells (0.14 μmol/mg cells/min) but to significantly stabilise it (*Figure 2*, curve 1). The half-life period of $\triangle^1$-dehydrogenase is 140 days (160 transformations), net operation time is 20 days.

The drop in the $\triangle^1$-dehydrogenase activity determines the extension of the time necessary for a complete hydrocortisone transformation to prednisolone. It takes 2 h for immobilised cells of *A. globiformis* to carry out the first transformation and 4 h to perform the two-hundredth. The activity drops because of the number of viable cells in the gel. After 60 transformations carried out during the first month, 75% cells remain viable. By the end of the second month the number diminishes to 35% and after 6 months (200 transformations) only 6% of viable cells remain.

Simultaneously with investigations on the enzyme activity and viability of cells, we have carried out electron microscopic examinations of IC. Samples have been taken every 5 days during the first month, then every other fortnight (two weeks). From *Figure 3a,b* it can be seen that the gel has a large-pore structure. Entrapment within the gel hardly disturbs the cell ultrastructure, and cells are uniformly distributed. However cells with a disturbed ultrastructure do accumulate gradually within the gel with lysed cells occurring deep in granules. In the subsurface layers, cells remain intact for a longer time, and moreover their number even increases, indicating the propagation of cells within the gel. Probably immobilised cells grow on buffer solution at the expense of lysis products. By the end of the experiment, in the subsurface layers of granules there are mainly intact cells. Inside the granules all the cells are lysed (22).

It is logical to assume that the level of the enzymic activity of the immobilised

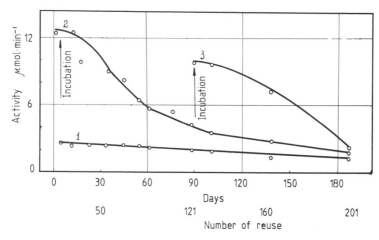

Figure 2. Changes in the 3-ketosteroid-$\triangle^1$-dehydrogenase activity during repeated batch-wise transformations of hydrocortisone. 1, granules not incubated in nutrient medium; 2, granules incubated once in nutrient medium; 3, granules incubated twice in nutrient medium.

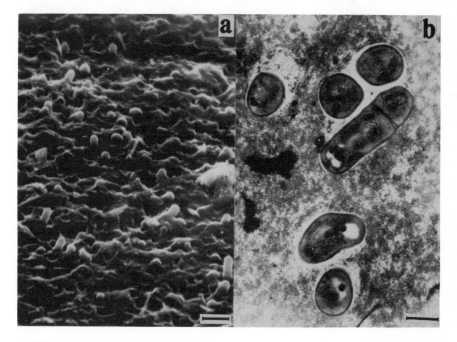

Figure 3. Cell distribution in gel and ultrastructure. (**a**) Gel granule surface (scale bar 1 μm); (**b**) cell ultrastructure after immobilisation (scale bar 0.5 μm).

system can be controlled by changing the intensity of cell propagation within the gel. To do this granules should be incubated in the nutrient medium supplied with an inducer. We have studied the regularity of the cell propagation within the gel, cell distribution and changes in the cell ultrastructure after batch-wise incubations of granules in the nutrient medium and in the course of repeated batch-

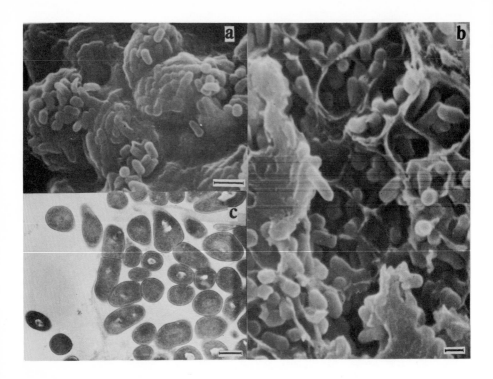

Figure 4. Polyacrylamide granules with cells after incubation in the nutrient medium. (**a**) Subsurface cell colony (scale bar 1 μm); (**b**) colonies inside granule (scale bar 0.5 μm); (**c**) cell ultrastructure after incubation in nutrient medium (scale bar 0.5 μm).

wise transformations in the buffer medium. Particular attention should be paid to the interrelationship between morpho-physiological changes in cells and their 3-ketosteroid-$\triangle^1$-dehydrogenase activity.

Granules containing the entrapped cells produced under sterile conditions are incubated for 24 h in the nutrient medium under conditions similar to those used for cultivation. After incubation, granules are filtered, washed repeatedly with sterile phosphate buffer or physiological solution and used for further transformations or kept in a refrigerator. From *Figure 2* (curve 2) it can be seen that after incubation in the nutrient medium there is a 5-fold increase in activity from 2.5 to 12.6 μmol/min. In the course of repeat transformations the $\triangle^1$-dehydrogenase activity decreases gradually. After 200 transformations in 6 months, 72% of the initial activity remains. In order to maintain a high level of $\triangle^1$-dehydrogenase activity of incubated IC, granules should be incubated in the nutrient medium after 3 months of continued use (*Figure 2, curve 3*).

We have also studied the distribution and ultrastructural changes in cells entrapped in granules after incubation in nutrient medium. After a single incubation of granules with cells in a full-value nutrient medium (1 day), the gel becomes filled with the cells of a new population (*Figure 4a − c*). Ultrathin sections of granules show that cells impregnated within the gel are morphologically intact, and dividing cells are evident (*Figure 4c*). In the course of subsequent transform-

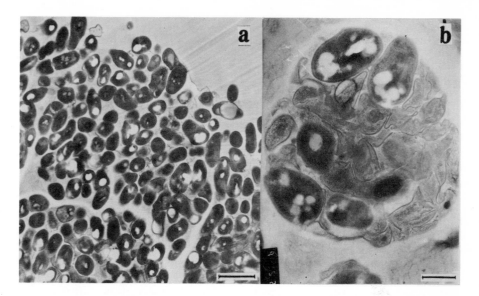

Figure 5. Cell ultrastructure in a granule of gel during batch-wise transformation. (a) On the surface of a granule layer after 6 months (scale bar 1 μm); (b) in the deep layer of a granule after 3 months (scale bar 0.5 μm).

ations over 6 months, intact cells prevail in the surface layers of granules. (*Figure 5a*).

In deeper layers, the cell lysis is more intensive, evidently due to the establishment of extreme anaerobic conditions. Ultrathin sections of middle layers of granules point to a complete cell lysis 2 − 3 months after immobilisation (*Figure 5b*). After the repeat incubation the gel is filled with cells of a new population, which is, however, largely heterogenous and consists of not only intact cells but also partially and completely lysed ones.

To maintain the viability of a cell population in gel it is advisable to carry out batch-wise incubations of granules in rich nutrient media followed by transformations in buffer media.

2.2.4 $\triangle^1$-Dehydrogenation under Continuous Conditions of Transformation (Type 1 and 2 Reactors)

(i) Place granules containing entrapped cells of *A. globiformis* (at a concentration of 1.9 or 11 mg/ml granules) in thermostatted glass columns 2 cm in diameter (see *Figure 1*).

(ii) Using 18 ml of granules in the column pass a solution of hydrocortisone (0.1 − 0.2 mg/ml) in 10 mM sodium phosphate buffer, pH 7.2, through the column filled with granules using a peristaltic pump ('LKB').

Under stationary conditions, a steady gradient of a substrate and a product is established which depends on the substrate flow-rate expressed as:

$$SV = \frac{1 \text{ ml solution}}{1 \text{ ml granules/h}} = 1 \text{ h}^{-1}$$

Or in units of linear rate: ml/h

One of the basic factors determining the selectivity of this process is the cell concentration in the carrier. Granules containing no more than $0.2 - 1.9$ mg cells/ml gel at a substrate flow-rate of SV ≥ 2.0 selectively accomplish $\triangle^1$-dehydrogenation of hydrocortisone and a number of other steroids: cortisone, Reichstein's substance S, methyltestosterone. As the cell concentration in the granules increases the selectivity of the process is disturbed, e.g., the 20β-hydroxy derivative of hydrocortisone is also produced (22).

Under conditions of selective $\triangle^1$-dehydrogenation, the $\triangle^1$-dehydro derivative yield is $90 - 95\%$ at a concentration of initial substrate of no more than 0.1 mg/ml. If the concentration of hydrocortisone is increased to 0.2 mg/ml, the yield of a $\triangle^1$-dehydrogenated product considerably decreases. In this case only $50 - 60\%$ prednisolone is accumulated in the reaction medium. The cells entrapped in polyacrylamide gel granules produce such quantities of prednisolone within wide temperature $(10 - 50°C)$ and pH $(6.0 - 9.0)$ ranges. Both free cells and those immobilised in gel under batch-wise conditions show maximal activity at $37°C$ and pH $7.0 - 8.0$.

The process of $\triangle^1$-dehydrogenation is stable at neutral and alkaline pH $(7.0 - 9.0)$ and temperature $(20 - 25°C)$. It should be noted that the prednisolone yield in the reaction medium is maintained constant, irrespective of the substrate flow-rate within $5 - 300$ ml/h (SV $=0.12 - 30.0$/h). The process is stable at low SV values. At SV $= 1.05 - 2.1$/h the half-life of the cells entrapped in gel under optimal conditions $(1 - 2$ min at $4°C)$ is 25 days. The disturbances of immobilisation conditions (elevation of temperature and extension of polymerisation contact) reduce the stability of $\triangle^1$-dehydrogenation. The half-life of the process carried out by the cells entrapped in gel during 15 min at $10 - 12°C$ is $6 - 7$ days (21).

As has already been mentioned, a substantial disadvantage of $\triangle^1$-dehydrogenation in continuous transformations (type 1 reactor) is the incomplete conversion of hydrocortisone to prednisolone caused by oxygen deficiency in the reaction medium. Additional aeration of granules in the column (type 2 reactor) produces a stimulating effect. With the upstream substrate flow (SV $=2.0$/h) and the air-bubbling through the column, hydrocortisone is totally transformed into prednisolone (IC concentration 1.9 mg/ml granules). Under these transformation conditions the use of microcrystalline hydrocortisone at a concentration of 1 mg/ml is also possible. At a flow-rate of 4.0/h, 55% prednisolone is produced. After recirculation of the effluent through the column, the prednisolone yield increases up to $88 - 90\%$.

A decrease in the enzymic activity of immobilised cells observed in the course of continuous transformation is accompanied by a drop in their respiration activity and viability as well as by a change of their ultrastructure. As revealed by transmission electron microscopy, after entrapment into the gel and during the first days of continuous transformation, the bulk of cells within the gel remain morphologically intact (*Figure 3b*). By the fifth day microcolonies consisting of a large number of bacterial cells are found on the surface of granules. Growing microcolonies are often situated close to cell-free clusters (*Figure 6a*).

Examination of paraffin sections of the granules confirms the cell propagation

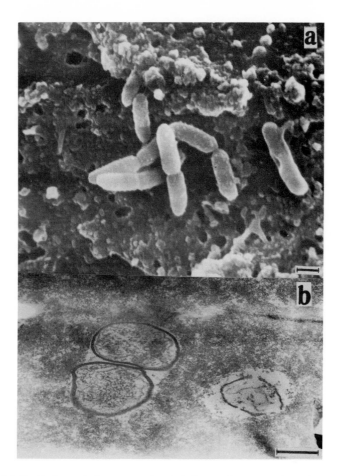

Figure 6. Cell distribution and ultrastructure in gel during continuous transformation. (a) Development of microcolonies on the gel surface (5 days) (scale bar 0.5 μm); (b) cell lysis in gel (25 – 30 days) (scale bar 0.5 μm).

not only on their surface but also inside them. In the subsurface layers of granules, separate regions are found with clusters of intact or dividing bacteria. We believe that IC growth should be considered as a manifestation of cryptogenic growth in buffer medium. At the expense of the cell lysis products the cryptogenic growth described has a receding character, therefore the number of lysed cells in the IC population prevails as the duration of transformation increases. By day 25 – 30 practically all the population consists of lysed cells (*Figure 6b*). This period is characterised also by an irreversible drop in the $\triangle^1$-dehydrogenase activity.

To stabilise $\triangle^1$-dehydrogenation, under continuous transformation conditions it will probably be expedient to use the granules pre-incubated in a nutrient medium in the presence of an inducer. Periodical incubation of granules in the nutrient medium during a decrease in activity makes it possible to maintain a viable popula-tion with a high degree of $\triangle^1$-dehydrogenase activity in the gel.

2.3 Entrapment in Calcium Alginate

Immobilisation of *A. globiformis* cells in 2% (w/v) calcium-alginate is conducted by the standard method (23) – see also Chapter 3.

(i)   Mix 100 mg of cells in 6 ml of physiological solution with 14 ml of 3% (w/v) sodium alginate solution.

(ii)   Add the mixture obtained dropwise, using a syringe and needle, into a stirred 10 mM calcium chloride solution.

(iii)   Wash the granules or fibres formed in physiological solution and use for transformation.

It has been established that the rate of hydrocortisone transformation by *A. globiformis* cells entrapped into calcium alginate is determined to a considerable degree by the shape of the carrier and the size of the granules. The highest rate is characteristic of cells entrapped into a fibre-like carrier (diameter of fibre – 1 mm). The transformation rate obtained by cells entrapped in gel granules (4 – 5 mm) is 30 – 40% lower. One of the factors determining the lower IC activity in gel granules is probably a limited rate of substrate diffusion into cells. The specific $\triangle^1$-dehydrogenase activity of cells entrapped into the fibres of calcium alginate corresponds to the activity of free cells and is 0.14 μmol/mg cells/min.

2.4 Entrapment in $\varkappa$-Carrageenan

To immobilise *A. globiformis* cells, use 4% (w/v) carrageenan. Carry out the immobilisation in the following way.

(i)   Mix 0.8 g of $\varkappa$-carrageenan dry powder with 17 ml of 0.9% (w/v) sodium chloride, and autoclave it at 112°C for 30 min.

(ii)   Cool the melted carrageenan down to 45 – 50°C, and rapidly mix with 3 ml of 0.9% (w/v) sodium chloride solution containing 100 mg of cells (dry weight).

(iii)   Introduce the resulting mixture dropwise (using a syringe) into a stirred cooled 0.3 M potassium chloride solution to obtain granules 3 – 4 mm in diameter.

(iv)   Obtain membranes (10 x 0.6 cm) by pouring the mixture of carrageenan and cells into a Petri dish and cut the membranes into cubes.

(v)   Store the immobilised cells at 4°C for 15 h in 0.3 M potassium chloride solution and use them to transform hydrocortisone to prednisolone.

Experimental data indicate that the enzymic activity of cells entrapped into this carrier is lower than that of free cells. When cells are entrapped in gel granules (diameter 4 mm) the $\triangle^1$-dehydrogenase activity decreases by 60%. It has been established that the basic factors determining the decrease in the enzymic activity of gel-entrapped cells are high temperature (45 – 50°C) and duration of immobilisation.

Preservation of $\triangle^1$-dehydrogenase activity of *A. globiformis* cells can be achieved using the following modification of the immobilisation procedure.

(i)   Pour 10 ml of the autoclave-melted carrageenan [0.8 g in 20 ml of 0.9% (w/v) NaCl] into a Petri dish.

(ii) After gel formation apply to the gel surface a concentrated cell suspension [100 mg (dry weight) in 0.8 ml of 10 mM sodium phosphate buffer pH 7.2] with 0.17 ml of ethanol containing 1.2 mg of hydrocortisone.

(iii) Air-dry the cell suspension uniformly distributed on the gel surface, for 15 – 20 min.

(iv) Apply 10 ml of melted carrageenan cooled down to 45 – 50°C to the gel surface resulting in cells confined between two layers of gel membranes.

(v) Store for 15 h in 0.9 M potassium chloride solution at 4°C and use the membranes for transformation.

This modified immobilisation technique makes it possible to reduce the time of the high temperature influence on cells. A specific $\triangle^1$-dehydrogenase activity of cells immobilised in this way is 0.11 μmol/mg cells/min which amounts to 78% of the free cell activity. The enzymic activity of the cells entrapped into the gel in the presence of the substrate is 10 – 15% higher than in its absence.

However, under batch-wise transformation conditions the activity is not stable, the half-life being 10 days. We believe that a rapid drop in the activity is determined by the wash-out of cells from the carrier. The treatment of the latter by the linking agent glutaraldehyde – even for a short time (2 – 3 min) – inhibits the activity almost completely.

2.5 Entrapment in Polyvinyl Alcohol and Other Artificial Membranes

Polyvinyl alcohol (PVA), cellulose diacetate, cellulose triacetate, polystyrene and polyacrylonitrile can be used for the immobilisation of cells.

(i) Prepare a 10% (w/v) PVA solution in distilled water, heating until it completely dissolves, and then cool the solution down to room temperature.

(ii) Add 10 ml of the PVA solution and a suspension of A. globiformis cells in water (16 – 32 mg/ml) into a 50 ml beaker, thoroughly mix until the suspension is homogeneous and pour into a Petri dish (diameter 10.5 cm) to achieve a uniform distribution.

(ii) Subject the mixture to u.v. irradiation (lamp PRK – 2 m) for between 15 and 90 min.

(iv) After drying, separate the film formed and cut into 100 – 200 mm² fragments and wash these fragments for 2 h with cool tap water.

(v) Carry out the transformation in Erlenmeyer's 250 ml flasks containing 25 ml of the reaction medium. The final concentration of hydrocortisone is 1 mg/ml.

To entrap cells in the other carriers mentioned, the carriers are dissolved in a suitable organic solvent (benzol, acetone, etc.), mixed with cells and air dried. The membranes obtained are fragmented and used for transformation. It has been established that the $\triangle^1$-dehydrogenase activity is retained only during entrapment into PVA membranes. The activity of IC rises as the cell biomass in membranes increases. At a cell concentration of 8 – 16 mg/ml medium, simultaneously with the basic process of $\triangle^1$-dehydrogenation, 20β-reduction and accumulation of the 20β-reduced prednisolone derivative occurs in the medium. Destruction of steroids is not observed.

2.6 Entrapment in Photocross-linkable Polymers

A. globiformis cells have been entrapped in photocross-linkable polymers according to the method of Sonomoto *et al.* (24). The following methodology may be followed.

(i)   Melt polyethylene glycol methacrylate (0.5 g) at 40°C.

(ii)   Mix ENT-101 (0.5 g) without heating with 0.04 − 0.2 g cells (dry weight) in 1.5 ml of 10 mM sodium phosphate buffer pH 7.2.

(iii)   Mix all components thoroughly, add 5 mg of crystalline benzoic ether, and thoroughly re-mix.

(iv)   Uniformly distribute the homogeneous suspension on a quartz glass, the area of the suspension is provided by a stainless steel ring (i.d. 60 mm, wall height 0.5 mm).

(v)   Cover the suspension carefully with a cellophane sheet avoiding air bubbles and then with glass.

(vi)   Irradiate the resulting suspension using u.v. light.

(vii)   Cut the film into squares (5 x 5 mm), wash and use them for transformation upon completion of polymerisation (in 3 − 5 min).

N.B. The distance between the u.v. source and the lower quartz glass is 200 mm. It is important that the glass is kept cool during irradiation and subsequent polymerisation.

Studies on the IC viability have shown that the bulk of cells isolated from the polymer obtained form colonies on the solid nutrient medium; as evidenced by electron microscopy, *A. globiformis* cells propagate in the given polymers. Living *A. globiformis* cells immobilised in these membranes carry out $\triangle^1$-dehydrogenation of hydrocortisone under batch-wise conditions (*Table 4*).

By changing the ratio of substrate to biomass, it is possible to inhibit the side reaction in order to obtain prednisolone practically without the admixture of 20β-reduced prednisolone. The $\triangle^1$-dehydrogenase activity of IC is sufficiently stable; after a 3 week storage of membranes at 4°C the activity remains unaltered.

During batch-wise transformations of hydrocortisone, the half-life of $\triangle^1$-dehydrogenase is 30 days (net operation time 20 days). The time required for 90%

Table 4. Transformation of Hydrocortisone by *A. globiformis* 193 cells Entrapped into Photocross-linkable Polymers.

Conditions		Transformation products, %					
Immobilised cells (mg/ml)	Hydrocorti- sone (mg/ml)	After 2 h			After 20 h		
		F^a	$\triangle F^b$	$20\beta\triangle F^c$	F	$\triangle F$	$20\beta\triangle F$
0.1	0.2	5	85	5	−	50	50
0.1	1.0	60	40	−	10	90	−
0.02	0.2	40	60	−	−	80	20
0.04	1.0	60	40	−	−	80	10

[a]F, hydrocortisone
[b]F, prednisolone
[c]$20\beta\triangle F$, 20β-hydroxyprednisolone derivative.

transformation of hydrocortisone to prednisolone increases from 2 to 48 h. Incubation of membranes in the growth nutrient medium in the presence of an inducer almost doubles the biocatalyst activity and considerably stabilises the process. The increase in the activity is determined by the cell growth in subsurface membrane layers.

2.7 Entrapment in Agar Gel and Linked Protein Membranes

2.7.1 *Agar Gel*

(i)   Mix a suspension of cells (100 mg/ml buffer) with 3% agar-agar heated up to $45 - 50°C$.
(ii) Cool the mixture down to 20°C and fragment the gel produced into granules by pressing through a sieve $(20 - 30$ mesh).
(iii) Wash the granules with 10 mM sodium phosphate buffer, pH 7.2 and use them for transformation under batch-wise and continuous conditions.

2.7.2 *Protein Membranes*

(i)   Dissolve 75 mg serum albumin and 0.12 g starch dialdehyde in 6 ml of 10 mM sodium phosphate buffer pH 7.2.
(ii) Add *A. globiformis* cells (25 mg) to the solution, thoroughly mix, pour into a Petri dish, and dry at 4°C.
(iii) Cut the membranes, wash with buffer and use for transformation.

The specific $\triangle^1$-dehydrogenase activity of *A. globiformis* cells entrapped in the carriers equals that of the free culture and amounts to 0.14 μmol/mg cells/min.

2.8 Adsorption

Adsorption is the very first known and simplest method of microbial cell immobilisation (see Chapter 1). It excludes the use of compounds toxic to cells but makes it possible to maintain the viability of microorganisms. However, the technique possesses considerable drawbacks: inadequate strength of cell retention and limited quantity of biomass adsorbed by a carrier unit.

2.8.1 *Adsorption on Cellulose*

To immobilise *A. globiformis* cells, various kinds of cellulose and large-pore ceramics can be used as carriers (*Table 5*).
Cells are immobilised in the following way.

(i)   To a cellulose suspension (7 g) in buffer $(150 - 20$ ml), add a bacterial suspension $(5 - 10$ ml) containing 100 mg of cells.
(ii) Shake at 200 r.p.m. for $20 - 30$ min at $20 - 22°C$ and leave overnight at 4°C.
(iii) To separate free cells, decant the carrier 10 times using 10 mM sodium phosphate buffer, pH 7.2 or centrifuge the carrier for 15 min at 1500 g and wash with buffer.

In the case of continuous transformation, unadsorbed cells can be separated after filling up the column. The extent of wash-out and cell desorption is deter-

Table 5. Basic Structural Characteristics of Carriers: Quantity and Activity of Adsorbed *A. globiformis* 193 Cells.

| Carrier | Characteristics of Carrier | | | | | | Quantity of immobilised cells (mg/g) | Activity | |
	Salt/kaolin (g/g)	Volume density (g/cm³)	Pore volume (cm³/g)	Specific surface (m²/g)	Average pore diameter (μm)	Granule size (mm)		μmol/min/g of carrier	μmol/min/mg of cells
CC-1	1.25	0.69	0.55	1.2	1.8	1–2	4.0	0.55	0.138
CC-1						0.5–1	4.6	0.62	0.135
CC-1						0.25–0.5	4.6	0.69	0.15
CC-2	2.3	0.49	0.92	0.6	6	0.5–1	10.8	1.4	0.138
CC-3	3.3	0.35	1.5	0.5	12	1–2	10.0	1.42	0.142
CC-3						0.5–1	10.2	1.28	0.126
CC-4	3.32	0.37	1.39	0.5	11	1–2	9.9	1.39	0.140
CC-4						0.5–1	10.0	1.39	0.139
CC-4						0.25–0.5	11.0	1.51	0.137
CC-5	3.33	0.35	1.47	1.8	3.3	1–0.5	10.8	1.48	0.137
CC-5						0.25–0.5	11	1.54	0.140
CC-6	3.36	0.34	1.51	1.6	3.8	0.5–1	10.2	1.40	0.137
Spherochrome-2PG	–	0.49	1.93	0.9	4.0	0.25–0.3	7.4	1.03	0.139

mined by the decrease in the optical density (OD) of the eluate.

On addition of 14 − 16 mg cells/g carrier, the optimal cell adsorption reaches 6 mg/g. The carrier prepared in this way containing adsorbed cells can be used to perform both batch-wise and continuous transformations.

In the case of batch-wise transformations, the carrier with cells is suspended in 750 ml flasks in 200 ml of buffer containing a corresponding substrate solution. After each use, the carrier is washed 3 − 4 times with buffer and kept at 4°C until its re-use.

When conducting continuous transformations, 7 g of the carrier with cells are loaded into a column (2 x 18 cm) through which the hydrocortisone solution (0.2 mg/ml in 10 mM sodium phosphate buffer, pH 7.2) is passed.

A. globiformis cells immobilised by adsorption on various celluloses are capable of $\triangle^1$-dehydrogenation of steroids.

Under batch-wise conditions the rate of hydrocortisone oxidation by adsorbed cells is no different from that of free cells, and constitutes 0.11 − 0.12 μmol/min/mg cells. The oxygen consumption rate for free and adsorbed cells during oxidation of hydrocortisone is the same, being 50 − 65 nmol O_2/mg cells/min. The cell activity does not change during cell adsorption in cellulose (17).

In the absence of hydrocortisone, when the cell respiration is determined only by oxidation of endogenous substrate, the oxygen concentration decreases linearly along the column height, that is, the oxygen consumption rate at various flowrates (6 − 570 ml/h, SV = 0.5 − 51.2/h) is independent of its concentration in the medium within 250 − 8 μM, thus constituting a zero order magnitude. The oxygen consumption rate increases when hydrocortisone is introduced into the reaction medium (*Figures 7,8*).

Oxygen consumption by cells is measured with a polarograph LP-7 (Czechoslovakia) using the Clark's Teflon-coated platinum electrode. Oxygen concentration in the medium is taken as equal to 250 μM. 10 mM Tris phosphate buffer, pH 7.0, is used as a respiration medium. The temperature of measure-

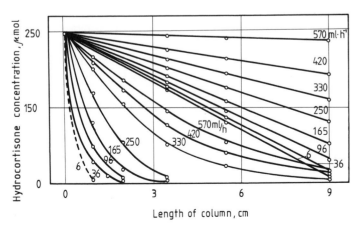

Figure 7. Profiles of oxygen concentration during endogenous respiration and $\triangle^1$-dehydrogenation of hydrocortisone by adsorbed cells of *A. globiformis* in the displacement column (2 cm in diameter).

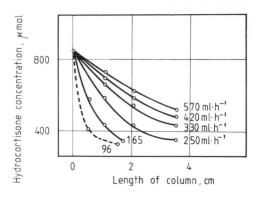

Figure 8. Profiles of hydrocortisone during its oxidation by adsorbed cells of *A. globiformis* in the displacement column (2 cm in diameter) at various solution flow-rates.

ment is $22 - 25 °C$, the final sample volume is 2 ml. To determine oxygen concentration in the column, a 0.5 ml flow cuvette is used.

The analysis of experimental data indicates that the main factor promising the efficient use of IC for continuous production of prednisolone in the column is the oxygen supply to the system.

At the optimal molar ratio of hydrocortisone to oxygen in the solution (2:1) a complete conversion can be attained at a flow-rate of $200 - 210$ ml/h (SV = 19.0, pH 7.2, $t = 20°C$; a column 2 x 3.5 cm in size contains 21 mg cells, 3.5 g cellulose; 0.18 mg/ml hydrocortisone). The half-life is 12 days at SV = 1.6/h.

A quantitative conversion of dissolved hydrocortisone to prednisolone is also registered at low flow-rates provided there is additional aeration (type 2 reactor). In the well-stirred and oxygenated reactor a complete conversion of hydrocortisone is feasible even at higher hydrocortisone concentrations of 0.5 and 1.0 mg/ml (17).

2.8.2 *Adsorption on Large-pore Ceramics*

The use of cells immobilised on a solid ceramic carrier is rather promising. Preservation of pore size and structure under various conditions (pressure, high flow-rates) protects microorganisms from a profound external effect. The above-mentioned carriers possess a high thermostability (1000°C) which make possible their regeneration just by air calcination.

Table 5 shows the data on the quantity of adsorbed *A. globiformis* cells depending on the specific surface and pore structure in various samples of large-pore ceramic carriers.

Carrier samples CC-2, CC-3 and CC-4 with average pore diameter $d_a \geq 6$ μm have surfaces similar in size (S = $0.5 - 0.6$ m²/g) and adsorb equal quantities of cells. Changing over to samples with smaller pores (CC-5 and CC-6) the sorbent surface increases 3-fold, but the quantity of cells adsorbed on the carriers remains the same.

At even smaller pore sizes (CC-1), a sharp decrease in the quantity of adsorbed cells is observed despite the fact that the S of this sample is twice as large as that of larger pore carriers. Therefore not all the surface of samples CC-5, CC-6 and CC-1 is available to cells.

The specific activity of IC in all the samples is approximately the same (0.13 − 0.15 μmol/mg/min) and is no different from that of free culture, thus indicating the absence of limitations for the diffusion of substrate into cells.

A. globiformis cells immobilised on ceramic carriers can be used repeatedly to perform transformation of hydrocortisone. Cells maintain their activity after 31 transformations (net operation time 15 days). During this time the initial rate of transformation remains constant consituting 1.4 μmol/g carrier/min (0.14 μmol/mg cells/min). By the 38th transformation (operation time 19 days) the initial reaction rate approximately halves, the time of a complete substrate conversion increases to 14 h and to 24 h by the 50th transformation.

The activity of immobilised cells on other carriers changes similarly. The least stable is the process catalysed by cells which are adsorbed on Spherochrome modified by pyrocarbon. The cells start losing their activity after the 20th transformation.

Under continuous transformation conditions, the process of $\triangle^1$-dehydrogenation is most stable if cells immobilised on the carrier CC-3, fraction 0.5 − 1 mm, are used. Passing a hydrocortisone solution (0.2 g/l) through the column (2 x 1.5 cm) at a rate of 50 ml/h, 50% conversion to prednisolone is maintained for 10 days (half-life 12 days). With the carrier CC-1, fraction 0.5 − 1 mm, the half life is 6 days.

Different stabilities are apparently due to variations in the surface structrue and pore size of the samples under study which determine the unequal strength of the cell retention on the carrier.

The decrease in the activity of adsorbed cells may result also from cell lysis; however the cell morphology has not be investigated in detail.

The possibility of increasing the quantity of adsorbed cells after a series of transformations by incubating in a nutrient medium appears to be rather important. As a result of incubation of the carrier with IC after 40 transformations their activity rises almost to the initial level. According to the quantitative determination of protein, the quantity of cells also increases up to its original value.

2.9 Covalent Binding

To carry out the covalent binding of *A. globiformis* cells, silica gel (40/100 μm) is used as a carrier. The gel is initially activated by $CrCl_3$ and $TiCl_3$ salts (see Chapter 2).

A. globiformis cells are immobilised in the following way.

(i)   For the activation by trivalent chrome salt Cr (III) suspend 2.5 g silica gel in 10 ml of $CrCl_3$ solution at various concentrations (30 − 45 mM) and add a cell suspension containing 40 mg cell/g carrier in 5 ml buffer (total volume is 15 ml in each case).

(ii) After 10 min at room temperature wash the carrier containing bound cells in 10 mM sodium phosphate buffer pH 7.2.

The cell quantity on the sorbent is determined from the difference between their contents in the suspension and wash water. Cells can also be bound after activating the support with trivalent titanium salt. However, if $CrCl_3$ is replaced by $TiCl_3$, cells lose their activity during immobilisation because the Ti (III) solution in hydrochloric acid used has a low pH. After activation of the carrier with titanium (III) the carrier must be washed extensively with distilled water prior to cell immobilisation.

The specific 3-ketosteroid-$\triangle^1$-dehydrogenase activity of cells immobilised on $CrCl_3$-activated silica gel corresponds to the activity of free cells (0.16 μmol/mg cells/min). Gradual desorption of cells occurs, as evidenced by microscopic examination of the reaction medium during repeated batch-wise transformations.

2.10 Summary of Immobilisation Techniques

To immobilise living *A. globiformis* cells with 3-ketosteroid-$\triangle^1$-dehydrogenase activity, a number of techniques have been described: entrapment in various gels including polyacrylamide, carrageenan, calcium alginate, agar, membranes of polyvinyl alcohol, photocross-linkable resins, protein membranes linked with glutaric dialdehyde; adsorption on various celluloses and large-pore ceramics; covalent binding with activated silica gel (*Tables 6,7*). Many of these techniques are modified to maintain the cell activity during production of the immobilised biocatalyst. *Table 6* shows variations in $\triangle^1$-dehydrogenase activity of IC under

Table 6. Comparative Characteristics of 3-ketosteroid-$\triangle^1$-dehydrogenase Activity of *A. globiformis* 193 Cells Immobilised by Various Techniques under Continuous Flow Conditions.

Immobilisation technique	SV/h	Half-life (days)
Entrapment into:		
Polyacrylamide gel	2.0	2.5
x-Carrageenan	—	—
Calcium alginate	—	—
Agar gel	1.7	14
Membranes of:		
PVA	1.0	12
Photocross-linkable resins	—	—
Protein membranes	—	—
Adsorption on:		
Powdered cellulose	1.6	12
DEAE-cellulose	1.6	15
Cellulose	1.6	10
Ceramic carrier fraction CC-3	5.3	22
0.5 − 1.0 mm	10	12
Covalent binding		
Free cells	1.4	6

Table 7. Transformation of Hydrocortisone by *A. globiformis* 193 Cells Immobilised by Various Methods under Batch-wise Conditions.

Immobilisation technique	Specific activity (μmol/mg cells/min)	Quantity of repeated transformations	Residual activity (% of initial activity)
Entrapment into:			
Polyacrylamide gel	0.14	200	44 after 6 months
x-Carrageenan	0.055	10	50 after 10 days
Calcium alginate	0.14	30	53 after 30 days
Agar gel	0.16		
Membranes from PVA		7	50 after 7 days
Photocross-linkable resins	0.14	20	30 after 2 months[a]
Protein membranes	0.14		
Adsorption on:			
Powdered cellulose	0.14 − 0.16		
Cellulose	0.16		
DEAE-cellulose	0.16		
Ceramic carrier, fraction CC-3 0.5 − 1.0 mm	0.15	50	40 after 50 days
Covalent binding:	0.16	2	50 after 3 days[a]
Free cells	0.16		

[a]Hydrocortisone concentration 0.2 g/l.

continuous transformation conditions. The stability is maximal for polyacrylamide gel-entrapped cells and those adsorbed on ceramic carriers. Under batch-wise conditions, $\triangle^1$-dehydrogenase activity is also highest during entrapment of cells into polyacrylamide gel (half-life 5 months, 144 transformations). Consequently, entrapment in polyacrylamide gel is one of the best methods for immobilisation of living *A. globiformis* cells with $\triangle^1$-dehydrogenase activity.

The other promising method of immobilisation of *A. globiformis* cells is by adsorption on large-pore ceramic carriers. In this case the stability of $\triangle^1$-dehydrogenase activity is lower than in polyacrylamide gel-entrapped cells but considerably higher compared with cells entrapped into other carriers.

It should be emphasised that during entrapment of *A. globiformis* cells in all the carriers except x-carrageenan gel, the $\triangle^1$-dehydrogenase activity is completely preserved and cells retain their viability.

3. STEROID TRANSFORMATIONS BY A. GLOBIFORMIS

3.1 $\triangle^1$-Hydrogenation

It is well known that free cells of *A. globiformis* perform the $\triangle^1$-reduction under anaerobic conditions and at a low redox potential. It is expected that $\triangle^1$-dehydrogenation and $\triangle^1$-hydrogenation are carried out by different enzymes (25). The cells retain their ability for $\triangle^1$-hydrogenation after their immobilisation on various carriers during continuous transformation (type 1 reactor).

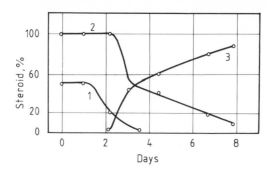

Figure 9. Effect of the quantity of entrapped cells on hydrogenation of prednisolone (SV=0.26/h). 1, hydrocortisone (biomass 1.9 mg/ml granules); 2, 20β-hydroxy derivative of hydrocortisone (biomass 11 mg/ml granules); 3, hydrocortisone (biomass 11 mg/ml granules).

To study the conditions of $\triangle^1$-hydrogenation, we have used prednisolone and the 20β-hdyroxy derivative of prednisolone as substrates at a concentration of 0.2 mg/ml, each of which is dissolved in ethanol in 10 mM sodium phosphate buffer (final concentration of ethanol, 2%). The final product of transformation of these substrates is the 20β-hydroxy derivative of hydrocortisone.

The data, characterising $\triangle^1$-hydrogenation of prednisolone, are presented in *Figure 9.* It is seen that the 3-ketosteroid-$\triangle^1$-reductase activity of IC is observed both at low (1.9 mg/ml granules) and high (11 mg/ml granules) concentrations of cells. At a concentration of 1.9 mg cells/ml granules (SV=0.26/h), 50% hydrocortisone is formed during the first day of transformation. By the second day the hydrocortisone yield decreases to 20%, there are no other products of transformation (*Figure 9,* curve 1). Therefore, at a low concentration of IC, the $\triangle^1$-reductase activity of a biocatalyst is not high and leaves the process of $\triangle^1$-dehydrogenation practically unaffected.

At the same substrate flow-rate and biomass, 11 mg/ml granules, the 20β-hydroxy derivative of hydrocortisone is the final and the only product of continuous transformation during the first 2 days (*Figure 9,* curve 2). Further, the yield of the 20β-hydroxy derivative of hydrocortisone decreases and, simultaneously, the hydrocortisone quantity increases (*Figure 9,* curve 3).

Similar results for the redox reactions have been obtained with *A. globiformis* cells immobilised on other carriers. The data obtained indicate that the production of the 20β-hydroxy derivative of hydrocortisone from prednisolone occurs under the consecutive action of two enzymes: 20β-hydroxysteroid dehydrogenase (20β-HSD) and $\triangle^1$-reductase of which $\triangle^1$-reductase is more active and stable than 20β-HSD (*Scheme 2*).

During continuous transformation, the 20β-hydroxy derivative of hydrocortisone is produced by IC *via* two possible routes. The $\triangle^1$-reductase activity of IC also falls within a wide range of substrate flow-rates. The process of hydrogenation is more stable at high concentrations of cells in the carriers, low substrate flow-rates, a temperature of 20°C and a pH of 7.0. pH and temperature optima of 1,2-reductase of immobilised and free cells are the same.

114

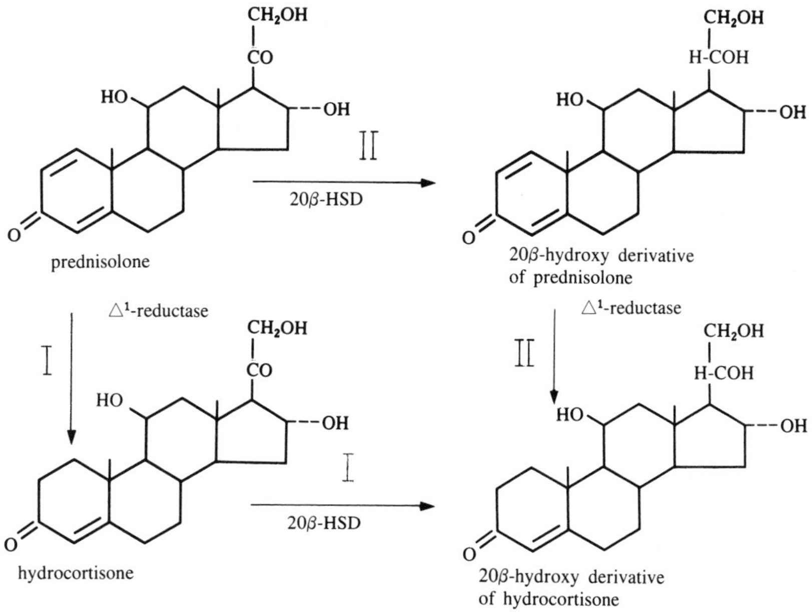

Scheme 2. Transformation of prednisolone to the 20β-hydroxy derivative of hydrocortisone.

3.2 20β-Reduction

Steroid substrates are reduced in position 20 of a steroid nucleus by 20β-HSD of *A. globiformis*.

After immobilisation of the culture on various carriers the activity of 20β-HSD remains unaltered. 20β-Reduction occurs selectively or with the other reactions, depending on aeration conditions.

3.2.1 *Batch-wise Transformation*

It has been shown that the 20β-reduced derivative of prednisolone is the final, but not the only, product of hydrocortisone and prednisolone transformation by IC using hydrocortisone as a substrate. The production of its 20β-hydroxy derivative is preceded by $\triangle^1$-dehydrogenation and prednisolone accumulation in the medium. The substrate is transformed according to *Scheme 3*.

The rate of accumulation of the indicated product is determined by the cell concentration in the carrier and is the same during transformation of both hydrocortisone and prednisolone. From this we may conclude that, under aerobic conditions, the $\triangle^1$-dehydrogenation is not a limiting stage (17).

The studies performed with free cells have shown that the process of 20β-reduction is stimulated by the presence of glucose. Transformation of steroids by free cells under aerobic conditions is performed as follows.

(i)   Incubate 25 mg of cells and hydrocortisone or prednisolone at a final concentration of 0.2 mg/ml (0.55 mM) in 250 ml flasks, containing 25 ml of triphosphate buffer, aerobically on a shaker at $180-220$ r.p.m.

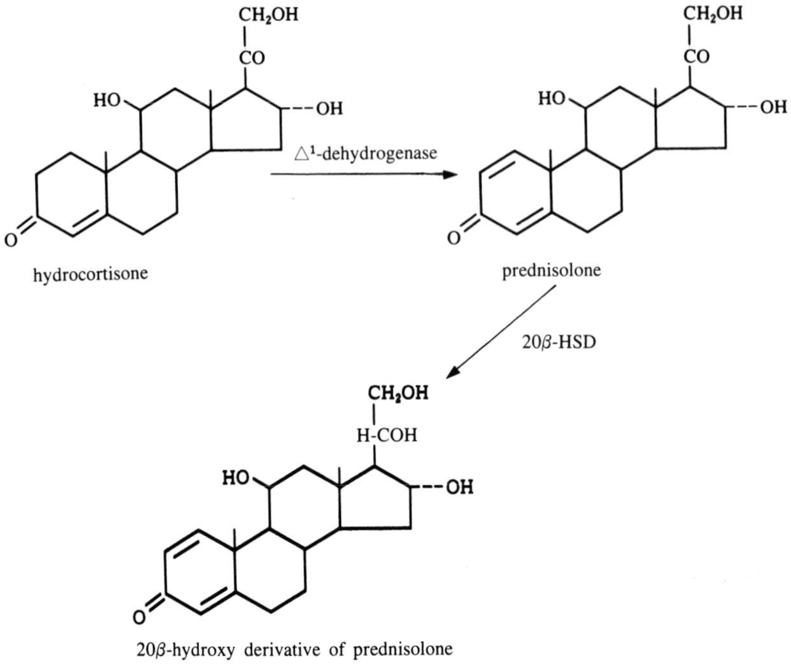

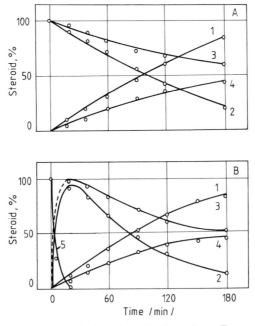

CH₂OH
|
CO

HO

--OH

$\triangle^1$-dehydrogenase

CH₂OH
|
CO

HO

--OH

O

hydrocortisone

O

prednisolone

20β-HSD

CH₂OH
|
H-COH

HO

--OH

O

20β-hydroxy derivative of prednisolone

Scheme 3. Transformation of hydrocortisone to the 20β-hydroxy derivative of prednisolone.

Figure 10. Effect of gluocose on prednisolone (**A**) and hydrocortisone (**B**) transformation by cells of *A. globiformis* 193. 1,3: prednosolone; 2,4: 20β-reduced derivative of prednisolone; 5: hydrocortisone; 4: transformation in the presence of glucose; 2: transformation in the absence of glucose.

(ii) Add glucose to the final concentration of 1% (w/v).

After 3 h of prednisolone transformation in the presence of glucose the content of prednisolone 20β-hydroxy derivative accounts for 85-90% of the initial level of prednisolone, i.e., 2−2.5 times higher than when produced in the absence of glucose (*Figure 10, A,B*). The increase of 20β-HSD activity in the presence of glucose is probably conditioned by the activation of metabolic processes which increases the level of the reduced pyridine nucleotide NADH which is a cofactor of 20β-HSD (17).

3.2.2 *Continuous Transformation*

Under continuous flow of substrate (type 1 reactor) IC transform a number of steroid hormones of the pregnane series (*Table 8*), producing their 20β-reduced derivatives.

Table 8. 20β-Reduction of Steroids by Cells Entrapped in Polyacrylamide Gel under Continuous Flow Conditions.

Substrate	Concentration of steroid (mg/ml)	Yield of 20β-hydroxy derivative	
		in 3 h	in 20 h
Hydrocortisone	0.2	100	100
Cortisone	0.2	100	100
Reichstein's	0.2	75	50
Substance 'S'	0.2	100	−
Prednisolone	0.2	100	100
Prednisone	0.1	100	−
△¹-derivative of			
Reichstein's substance 'S'	0.05	75	40

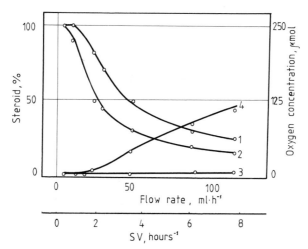

Figure 11. Effect of substrate flow-rate on the formation of the 20β-hydroxy derivative and the consumption of oxygen by adsorption on cellulose of cells of *A. globiformis* (column 21 ×6 cm, cell content 6 mg/g cellulose) **1.** substrate - hydrocortisone, **2.** substrate - prednisolone, **3.** oxygen concentration in medium at transformation of hydrocortisone, **4.** oxygen concentration in medium at transformation of prednisolone.

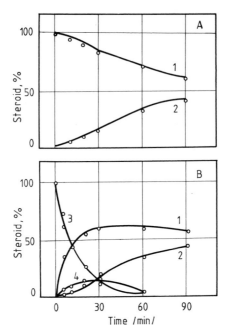

Figure 12. Transformation of hydrocortisone (**A**) and prednisolone (**B**) by *A. globiformis* cells under strictly anaerobic conditions. 1, hydrocortisone; 2, 20β-hydroxy derivative of hydrocortisone; 3, prednisolone; 4, 20β-hydroxy derivative of prednisolone.

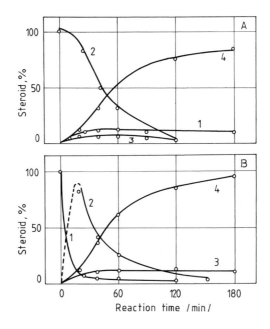

Figure 13. Transformation of hydrocortisone (**A**) and prednisolone (**B**) by *A. globiformis* cells under aerobic conditions in the presence of cyanide. 1, prednisolone; 2, hydrocortisone; 3, prednisolone; 4, 20β-hydroxy derivative of hydrocortisone.

In the case of steroids possessing a $\triangle^1$-bond in ring A and 20β-hydroxy derivatives lacking a double bond in the $\triangle^1$-position, there is no oxygen in the column eluate (*Figure 11*).

It has been found that, under anaerobic conditions or in the presence of cyanide in the incubation medium during transformation of prednisolone, it is hydrocortisone which accumulates first which is then gradually transformed into its 20β-hydroxy derivative (*Figures 12,13*) (17). Therefore, the basic pathway of hydrocortisone 20β-hydroxy derivative production includes a prednisolone reduction to hydrocortisone catalysed by 3-ketosteroid-$\triangle^1$-reductase followed by the direct reduction of the 20β-keto group of the latter under the action of 20β-HSD (*Scheme 4*).

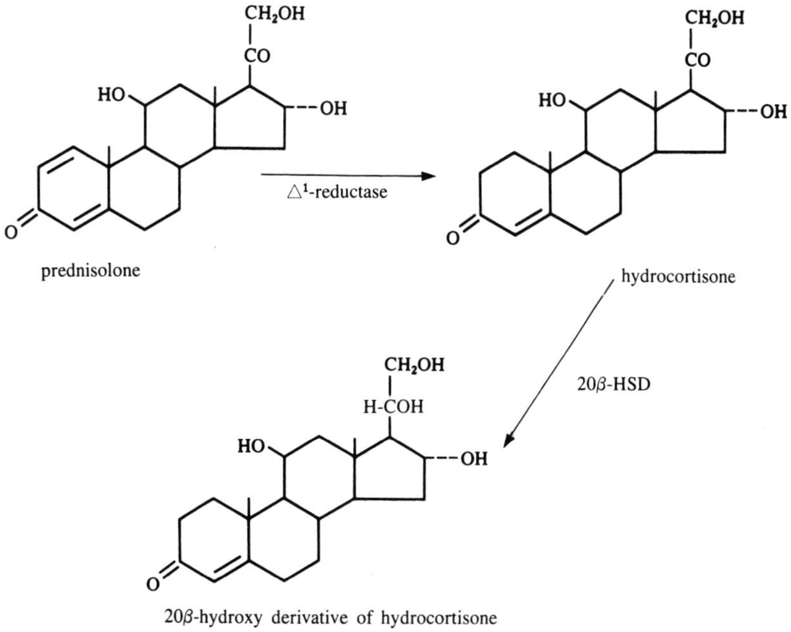

Scheme 4. Transformation of prednisolone to the 20β-hydroxy derivative of hydrocortisone.

Such a sequence of prednisolone transformation is conditioned evidently by the fact that the activity of 3-ketosteroid-$\triangle^1$-reductase is considerably higher than that of 20β-HSD (17).

Hydrocortisone 20β-hydroxy derivative is produced from hydrocortisone by two possible routes (*Scheme 5*). The activity of $\triangle^1$-dehydrogenase is inhibited by cyanide and by strictly anaerobic conditions. Under inhibition of 3-ketosteroid-$\triangle^1$-dehydrogenase there is a direct reduction of the hydrocortisone 20-keto group. Analysis of the results obtained indicates that, under continuous conditions, hydrocortisone transformation by IC occurs *via* two pathways (*Scheme 5*). In the upper part of the column, in the presence of oxygen, hydrocortisone 20β-hydroxy derivative is produced by the multi-enzyme system (pathway I). Due to high $\triangle^1$-dehydrogenase activity in the presence of oxygen, all hydrocortisone

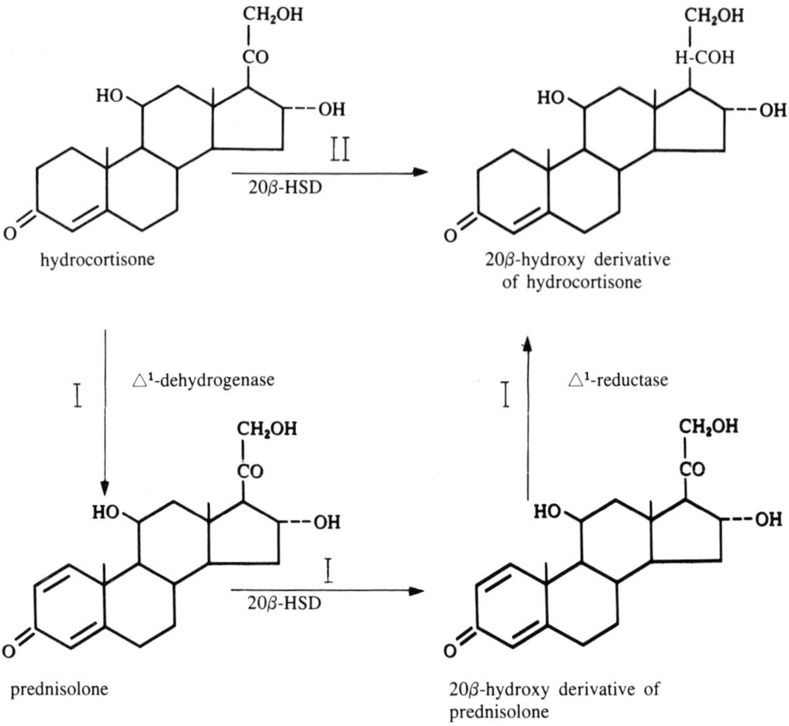

Scheme 5. Transformation of hydrocortisone to the 20β-hydroxy derivative of hydrocortisone.

Table 9. Stability of Hydrocortisone 20β-Reduction at Different Specific Flow-rates.

Inducer	SV/h	Time (days)[a]
Cortisone acetate	0.07	5
	0.14	4
	0.52	2
	0.7	34 − 36
	1.05	1
	2.1	8 − 10
Cortisone acetate + progesterone	0.07 − 0.14	9
	0.25 − 0.52	5

[a]100% conversion of hydrocortisone to its 20β-hydroxy derivative.

is rapidly oxidised to prednisolone and direct 20β-reduction of hydrocortisone is unlikely. Under anaerobic conditions in the deeper layers of the carrier in the column the direct reduction of hydrocortisone occurs (pathway II) (17).

The basic factors which determine the stability of the 20β-reduction process under optimal conditions (20 − 22°C, pH 7.0) are the concentration of IC in the carrier and the substrate flow-rate. Stability increases with an increase in the amount of biomass and with a decrease in the flow-rate.

The data on the stability of hydrocortisone 20β-reduction performed by the

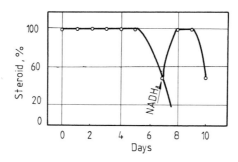

Figure 14. Effect of exogenous NADH on 20β-reduction (IC concentration is 11 mg/ml gel, SV=0.65/h).

polyacrylamide-entrapped cells are presented in *Table 9*.

As is known, the enzymic activity and direction of transformation processes may be regulated by various means, primarily by selective induction of certain enzymes with the help of steroid compounds. 20β-HSD belongs to the group of inducible enzymes. Progesterone is the most effective inducer (26). The culture is grown in conditions of maximal synthesis of the enzyme as follows.

(i)   After 24 h growth (*Table 1*) remove the cells by centrifugation and wash them with buffer.

(ii)   Place 2 g cells (dry weight) in a 750 ml flask containing 150 ml 10 mM sodium phosphate buffer, pH 7.2, supplemented with 0.25% glucose and corn-steep extract and progesterone dissolved in ethanol at a concentration of 10 mg%.

(iii) Incubate the cells on a shaker at 180 r.p.m., 37°C for 5 h.

(iv)   Immobilise the cells in polyacrylamide gel and use for transformation.

As is seen in *Table 9*, the additional induction of the culture by progesterone provides an almost 2-fold rise in the stability of the 20β-reduction process.

20β-HSD is known to belong to the group of NADH-dependent enzymes. It may well be that the decrease in the yield of the hydrocortisone 20β-reduced derivative is determined by the exhaustion or washing out of the endogenous NADH.

In view of the above, we have examined the possibility of retaining 20β-HSD activity and stabilisation of the process in the presence of the exogenous cofactor. As is seen in *Figure 14*, NADH added on the 7th day of continuous transformation raises the yield of the product to the original level and facilitates the stabilisation of the process within several days. After the decline of 20β-HSD activity, prednisolone is the only product of hydrocortisone transformation.

Glucose also exerts a pronounced stabilising effect upon the 20β-reduction process. The addition of 1 mM glucose to the reaction medium results in an almost 2-fold increase in the half-life of the process performed by the cells adsorbed on cellulose, obviously due to the increase in NADH content in the cells (*Figure 15*). Thus, stabilisation of the 20β-reduction process is provided in several ways, of which the regeneration of the cofactor by the cell in the presence of glucose is the most economical way.

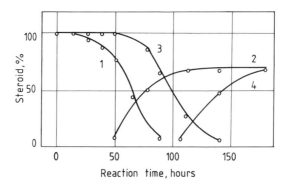

Figure 15. Effect of glucose on the dynamics of 20β-reduction and △¹-dehydrogenation during transformation of hydrocortisone by cells of *A. globiformis* adsorbed on cellulose under continuous column conditions (column 2 x 6 cm containing 40 mg cells). 1,2: hydrocortisone transformation in the absence of glucose; 3,4: hydrocortisone transformation in the presence of glucose; 1,3: 20β-reduced derivative of hydrocortisone; 2,4: prednisolone.

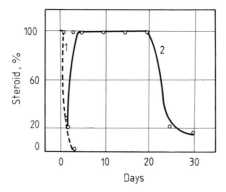

Figure 16. Hydrocortisone transformation in the presence of menadione. 1, hydrocortisone 20β-hydroxy derivative; 2, prednisolone (IC concentration 11 mg/ml granules).

Selective production of prednisolone, according to *Scheme 5*, is possible only after 20β-HSD activity declines. To inhibit 20β-HSD activity, the following procedures can be employed:

(i)   freeze-thawing of intact cells;
(ii)   thermal treatment of IC at 50°C;
(iii)   transformation in the presence of artificial electron acceptors.

The effect of exogenous quinones on 20β-reduction is tested by adding phenazine methosulphate and menadione to hydrocortisone-containing phosphate buffer. Menadione is introduced into the reaction medium with ethanol. The final concentration of ethanol is 1%. In this case, as is seen in *Figure 16*, at SV = 1.05/h, the 20β-hydroxy derivative is selectively produced within 24 h. On the 2nd day, along with hydrocortisone (30%), the 20β-hydroxy derivative of hydrocortisone (15 − 20%) is produced. Beginning with the 3rd day of continuous transformation, a quantitative yield of prednisolone is produced for the next 17 days (in the experiment the cells were entrapped in gel for 15 min at 15 − 18°C, non-optimal conditions of polymerisation). As is shown experimentally, *A. globiformis* IC remain capable of $\triangle^1$-dehydrogenation, 20β-reduction and $\triangle^1$-hydrogenation. The mode of enzymic processes is determined by the conditions of substrate transformation; for example, temperature, pH of the medium, substrate flow-rate, the presence of exogenous electron acceptors and cofactors and the concentration of cells in granules. The various combinations of these factors produces either $\triangle^1$-dehydrogenation, 20β-reduction or $\triangle^1$-reduction.

3.3 11α-, 11β-Hydroxylation

11α- and 11β-hydroxylation of steroid compounds is one of the most important reactions in the production of steroids. Microorganisms of various taxonomic groups, and most often fungi, are capable of microbial 11α- and 11β-hydroxylation of steroid compounds. Since the enzymes which catalyse the incorporation of hydroxyl groups into a steroid nucleus are extremely labile, sufficiently pure preparations have not been isolated so far. Most studies on hydroxylation of steroid compounds have been carried out using whole cells. Therefore, the mechanism of the reaction and the nature of the enzymes still needs further investigation. It is supposed that hydroxylation requires molecular oxygen, a system for regeneration of the cofactor and the cytochrome P_{450} oxidase system. This means that successful hydroxylation is possible only in a living cell.

The culture *Tieghemella orchidis* IBPhM F-233 can be used for hydroxylation of cortexolone and its 21 acetate in the 11α- and 11β-position (*Scheme 6*).

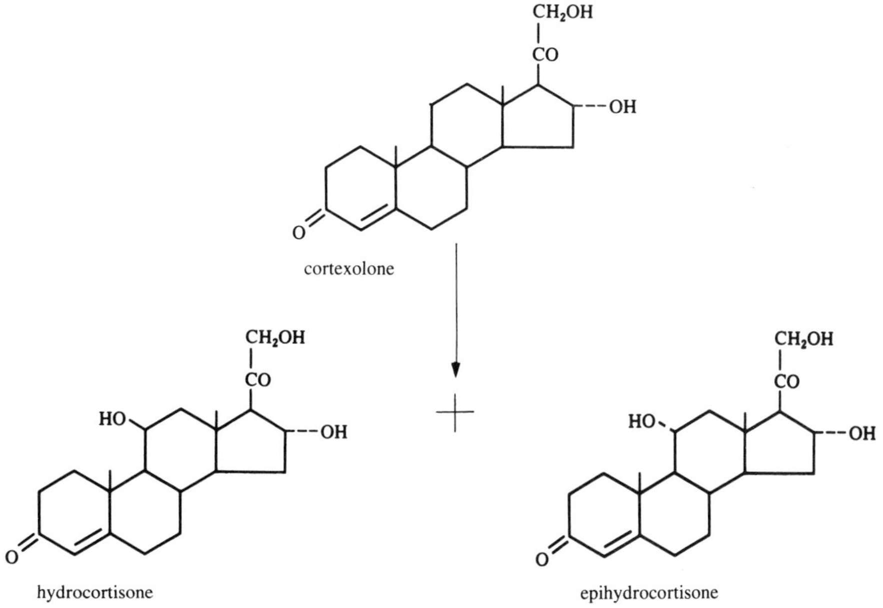

Scheme 6. Hydroxylation of cortexolone.

The hydroxylase activity of *T. orchidis* mycelium, entrapped in polyacrylamide gel under conditions optimal for *A. globiformis* cells, is close to that of the free mycelium, but is not stable. Mycelium entrapped in calcium alginate and photocross-linkable polymer is one quarter as active as the free mycelium, Inoculation in the nutrient medium gives an insignificant positive effect.

3.4 Asymmetric 17β-Reduction of 3-Metoxy-$\triangle^{1,3,5(10),9(11)}$ 8,14-Sekoestra-tetradion-14,17

The reaction of asymmetric 17β-reduction of 3-metoxy-$\triangle^{1,3,5(10),9(11)}$ 8,14-seko-estratetradion-14,17 (sekoketone) yielding sekoketol — an optically active compound — is of great practical importance. Sekoketol is a precursor in the synthesis of many steroid hormones. To carry out this reaction, free and immobilised cultures (IC) of *Saccharomyces cerevisiae* VKM u-488, which are highly specific and yield optically clear sekoketol (*Scheme 7*), can be used.

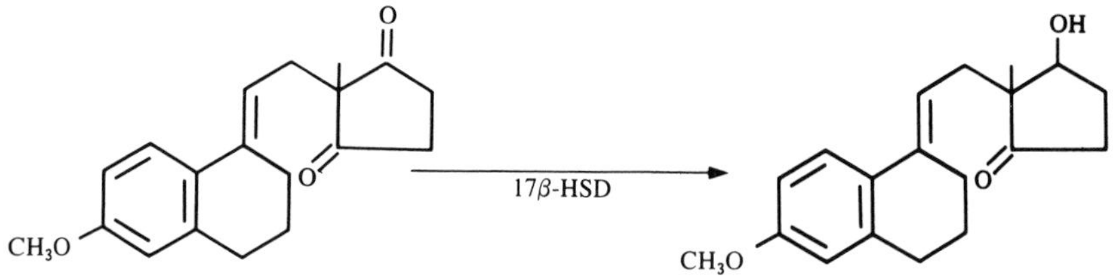

Scheme 7. Stereospecific 17β-reduction.

The reaction is catalysed by an inducible enzyme, 17β-hydroxysteroid dehydrogenase (17β-HSD) requiring reduced NAD as a cofactor (27). Yeasts are entrapped in polyacrylamide gel, or polyvinyl alcohol membranes, and are ad-sorbed on various carriers. The best result is obtained using polyacrylamide gel granules incubated in the nutrient medium. 17β-HSD activity in this case is equal to that of free yeast. After 45 transformations (sekoketone 1g/l), on day 45 the activity decreases by 20%.

4. REFERENCES

1   Koshcheyenko,K.A. (1981) *Prikl. Biokhim. Microbiol.*, **27**, 477.
2. Koshcheyenko,K.A. (1981) *Mikrobiologia*, **11**, 55.
3. Fukui,S. and Tanaka,A. (1982) *Annu. Rev. Microbiol.*, **36**, 145.
4. Kolot,F.B. (1982) *Process Biochem.*, **17**, 12.
5. Suzuki,S. and Karube,I. (1978) in *Enzyme Engineering*, Vol. **4**, Broun,G.B., Manecke,G. and Wingard,L.B., (eds.), Plenum Press, New York and London, p. 329.
6. Mosbach,K. and Larsson,P. (1970) *Biotechnol. Bioeng.*, **12**, 19.
7. Larsson,P.O., Ohlson,S. and Mosbach,K. (1978) in *Enzyme Engineering*, Vol. **4**, Broun,G.B., Manecke,G. and Wingard,L.B. (eds.), Plenum Press, New York and London, p. 317.
8. Ohlson,S., Larsson,P.O. and Mosbach,K. (1979) *Eur. J. Appl. Microbiol. Biotechnol.*, **7**, 103.
9. Malik,V.S. (1982) *Z. Allg. Mikrobiol.*, **4**, 261.

10. Atrat,P. (1982) *Z. Allg. Mikrobiol.*, **10**, 723.
11. Fukui,S., Omata,T., Yamane,T. and Tanaka,A. (1980) in *Enzyme Engineering*, Vol. **5**, Weetal, H.H. and Royer,G.P. (eds.), Plenum Press, New York and London, p. 347.
12. Fukui,S. and Tanaka,A. (1982) in *Enzyme Engineering*, Vol. **6**, Plenum Press, New York and London, p. 203.
13. Atrat,P., Huller,E. Hörhold,C., Buchar,M.I., Arinbasarova,A.Y. and Koshtschejenko,K.A. (1980) *Z. Allg. Mikrobiol.*, **20**, 159.
14. Atrat,P., Hörhold,C., Buchar,M.I. and Koschtschejenko,K.A. (1980) *Z. Allg. Mikrobiol.*, **20**, 239.
15. Skryabin,G.K., Zvyagintseva,I.S., Nazaruk,M.M. and Sokolova,L.V. (1965) *Dokl. Akad. Nauk USSR*, **161**, 472.
16. Skryabin,G.K., Zvyagintseva,I.S. and Nazaruk,M.M. (1966) *Prikl. Biokhim. Mikrobiol.*, **2**, 271.
17. Arinbasarova,A.Yu., Medentsev,A.G., Akimenko,V.K. and Koshcheyenko,K.A. (1985) *Biokhimiya*, **50**, 1,3.
18. Medentsev,A.G., Arinbasarova,A.Yu., Koshcheyenko,K.A. and Akimenko,V.K. (1983) *Biokhimiya*, **48**, 17.
19. Maurer,P. (1971) *Disc-electrophoresis, Theory and Practice of Electrophoresis in Polyacrylamide Gel*, published by Mir Publishers, Moscow, p. 13.
20. Freeman,A., Blank,T. and Aharowitz,Y. (1982) *Eur. J. Appl. Microbiol. Biotechnol.*, **14**, 13.
21. Koshcheyenko,K.A., Avramova,T.L. and Sukhodolskaya,G.V. (1981) *Izv. Akad. Nauk SSSR, Ser Biol.*, **2**, 174.
22. Koshcheyenko,K.A., Sukhodolskaya,G.V., Tyurin,V.S. and Skryabin,G.K. (1981) *Eur. J. Appl. Microbiol. Biotechnol.*, **12**, 161.
23. Kierstan,M. and Bucke,C. (1977) *Biotechnol. Bioeng.*, **19**, 387.
24. Sonomoto,K., Hoo,M.D.M., Tanaka,A. and Fukui,S. (1981) *J. Ferment. Technol.*, **59**, 465.
25. Bukhar,M.I. and Lestrovaya,N.N. (1970) *Biokhimiya*, **35**, 843.
26. Gotovtseva,N.N. and Korovkina,A.S. (1973) *Mikrobiologiya*, **52,3**, 434.
27. Koshcheyenko,K.A., Turkina,U.V. and Skryabin,G.K. (1983) *Enzyme Microb. Technol.*, **5**, 13.

CHAPTER 8

Immobilised Plant Cells: Preparation and Biosynthetic Capacity

PETER BRODELIUS

1. INTRODUCTION

Isolated plant cells can be cultivated *in vitro* in analogy to microbial cells in shaker flasks or in fermentors. The potential use of plant cell cultures for the production of biochemicals has long been recognised. Over 90% of approximately 30 000 known natural products can be found in higher plants. Cultivated plant cells are totipotent, i.e., the entire genetic information is available for expression, and therefore it should be possible to produce any compound found in the intact plant also in a culture of this species. At present, many commercially important compounds are isolated from tissue material obtained by field cultivation of plants or by collecting wild plants. These substances are used in the pharmaceutical, food and fragrance industries. They are of complex chemical structure and therefore difficult to produce by other means. The supply of certain important raw plant materials today is, or may be in the near future be, limited. It has become increasingly important to find and develop alternative resources. One such alternative resource is plant cell cultures and during the last decade major progress has been made in this area (1,2). Actually, the first commercial process for the production of a biochemical (shikonin) based on plant cell culture was recently introduced in Japan (3).

There are, however, some basic problems that have to be solved before plant cell cultures can be generally employed for the production of biochemicals on a commercial basis. Plant cells in culture differ very much from microorganisms and the advanced fermentation technology developed for the latter in most cases cannot be used directly for plant cell fermentations. Furthermore, special features of plant cells (e.g., slow growth, cell aggregation, low yields and genetic instabilities) make the utilisation on a large scale more complicated.

The advantages of immobilised biocatalysts are well recognised (4,5) and during recent years, we have been investigating if these advantages also apply to immobilised whole plant cells (6 – 8). This has been done by studying the biosynthesis of various compounds in model systems which were selected to cover different aspects of biosynthesis, i.e., synthesis *de novo* from a simple carbon source (e.g., sucrose), synthesis from added precursors or bioconversions (e.g., hydroxylations). For a preserved biosynthetic capacity, a viable cell preparation is of importance. Therefore, we have investigated various immobilisation procedures and studied the viability and biosynthetic capacity of these preparations.

2. IMMOBILISATION PROCEDURES FOR PLANT CELLS

Plant cells are relatively sensitive to changes in the environment and consequently only the mildest methods of immobilisation available may be employed. Of the basic techniques available we have found that the entrapment technique is most appropriate to use for large sensitive plant cells. There are also examples on the immobilisation of plant cells by adsorption and by covalent linkage. It must be possible to carry out the immobilisation under sterile conditions, while plant cells grow relatively slowly and therefore are very sensitive to infections by rapidly growing microbial cells. The material used for immobilisation must thus tolerate sterilisation (e.g., autoclaving).

2.1 Entrapment Techniques

There are several entrapment techniques available for the immobilisation of plant cells. The final cell concentration of the immobilised cell preparation may vary considerably [from a few percent to at least 50% (w/w)]. The optimal cell concentration has to be determined for each application and in the procedures described below a standard cell concentration of 20% (w/w) has been chosen.

2.1.1 *Alginate*

Calcium alginate is widely used for the immobilisation of various cell types (see Chapter 3) and it has also been used for the entrapment of various plant cells. In our first attempts to immobilise various cells we chose this polymer since the method is very mild, the polymer can be sterilised by autoclaving and the immobilisation can be reversed by addition of a calcium chelating agent (e.g., EDTA or citrate). In the initial studies we wanted to be able to reverse the immobilisation in order to study the cells after some time in the immobilised state.

The final concentration of alginate in the immobilised preparation may vary depending on the type of alginate used. The gel stability increases with increasing polymer concentration but at higher concentrations the alginate/cell suspension will be very viscous, which may cause problems in bead formation. Thus, a compromise between gel stability and ease of preparation has to be found. It should also be pointed out that, upon autoclaving, the viscosity of the sodium alginate solution is decreased due to partial hydrolysis of the polymer and consequently excessive autoclaving should be avoided.

Small batches of alginate-entrapped plant cells are conveniently prepared by the utilisation of disposable sterile plastic syringes. If a hypodermic needle is used to obtain smaller beads, care must be taken to avoid its clogging by larger cell aggregates. To reduce this kind of problem the cell suspension can be filtered through a sterilised screen (metal or nylon) of appropriate mesh size. Furthermore, the needle used should be cut to a length of a few millimetres to decrease the flow resistance. For details see *Table 1*.

The bead size can, to some extent, be controlled using different sized needles on the syringe. In this manner beads ranging from 2 to 4 mm in diameter may be prepared. If smaller beads are required, an experimental set up as shown in *Figure 1* can be used. The surface tension of the viscous cells/alginate solution is

Table 1. Small-scale Preparation of Alginate-entrapped Plant Cells.

1.	Sterilise a solution of sodium alginate (e.g., 2% protanal HF, Protan A/S, Norway) prepared in an appropriate medium by autoclaving (20 min, 121°C).
2.	Collect the plant cells by centrifugation or filtration and then suspend them (2.0 g wet weight) in the cooled alginate (8.0 g) in a sterile 25 ml beaker using a sterile spatula.
3.	Transfer the cell/alginate suspension into a disposable plastic syringe (10 ml) from which the piston has been removed.
4.	Place the syringe over an Erlenmeyer flask containing medium (50 ml) fortified wtih 50 mM CaCl$_2$ and allow the cell/alginate suspension to drop into the calcium-containing medium which is gently stirred.
5.	Leave the beads formed in this solution for 30 min and then collect and wash them with medium.
6.	Transfer the beads to an appropriate medium containing at least 5 mM CaCl$_2$.

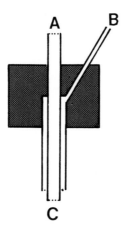

Figure 1. Schematic diagram of device used for the preparation of small alginate beads. (**A**) Inlet for alginate/cell suspension; (**B**) inlet for compressed air; (**C**) outlet for alginate/cell suspension with surrounding airstream.

overcome by a sterile air stream at the end of the needle. This procedure allows for the preparation of beads with a diameter of 0.2 − 1 mm. Normally, the immobilised cells are incubated in a batch procedure as described below.

During the 'ripening' of the beads, they will shrink to a certain extent due to exclusion of water as the calcium ions diffuse into the beads and bind the polymer chains more tightly together. This shrinking is different for different preparations of alginate and also dependent on polymer concentration. When quantitative experiments are required, the following procedure should be used.

(i)   Before addition of the alginate/cell suspension to the medium, weigh the flask containing the medium.

(ii) After addition of the suspension weigh it again.

(iii) Calculate the amount of alginate/cell suspension added and the wet weight of the beads after washing and filtration. The cell concentration (w/w) is then:

$$C = C_o \; \frac{\text{weight of alginate/cell suspension added}}{\text{weight of beads}}$$

where C = cell concentration in beads (w/w), C_o = cell concentration in alginate/cell suspension (w/w).

For the preparation of large quantities of alginate-entrapped plant cells a device can be constructed as shown schematically in *Figure 2*. The entrapment of plant cells within alginate using the device is described in *Table 2*.

2.1.2 *Kappa-carrageenan*

Kappa-carrageenan is a polysaccharide containing sulphate esters ($>20\%$ of the sugar moieties are sulphonated) and it is not soluble in cold water. It can, however, be dissolved by heating and, upon cooling, a gel is formed. The gelling temperature and the quality of the gel depend upon concentration of polymer as well as the amount and types of cations (e.g., K^+, NH_4^+, Ca^{2+} or Ba^{2+}) pre-

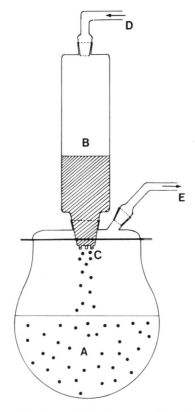

Figure 2. Schematic diagram of device used for the preparation of large amounts of alginate-entrapped plant cells. (**A**) Medium container with lid (total volume 1 litre); (**B**) reservoir for alginate/cell suspension (total volume 0.4 litre); (**C**) six tubes for drop formation (inner diameter 1 mm); (**D**) inlet for sterile air; (**E**) outlet for air.

Table 2. Large-scale Preparation of Alginate-entrapped Plant Cells.

1.	Autoclave the entire device (*Figure 2*) and weigh it with the calcium-containing medium (500 ml) in the medium container A.
2.	Mix the alginate (240 g) and plant cells (60 g wet weight) together and transfer them into the reservoir B.
3.	Apply a slight pressure of sterile air in order to press the viscous alginate/cell suspension through the six tubes C so that beads of uniform size are formed. Stir the medium gently during bead formation.
4.	Weigh the whole device and leave for 30 min.
5.	Collect the beads by filtration, wash, weigh and finally transfer them to an appropriate medium. From the weight data calculate the cell concentration within the beads as described in Section 2.1.1.

sent. For instance, to remain in a liquid form the kappa-carrageenan supplied by Sigma Chemical Co. (Type III) requires $45-50°C$ for a 5% solution in water, while a special preparation of carrageenan for the immobilisation of cells from FMC Corporation (NJAL 798) requires only $25-30°C$ at the same concentration. For immobilisation of sensitive plant cells, the latter type of carrageenan is the most appropriate to use. There are different procedures available for entrapment of cells in carrageenan as described in *Table 3*.

2.1.3 *Agar/agarose*

Agar and agarose (a purified preparation of agar) form gels upon cooling of a heated solution and the latter can be chemically modified (introducion of hydroxyethyl groups) in order to lower the gelling temperature. For the immobilisation of plant cells a gelling temperature of $25-30°C$ is convenient to use (e.g., Type VII from Sigma or SeaPlaque from FMC).

An advantage of agarose is that no counter ion is required for gel stability. The mechanical stability of an agarose gel is, however, lower than that of alginate or carrageenan at the same polymer concentration. Furthermore, the cost of agarose is considerably higher than for the other two polysaccharides.

The same procedures (B and C, *Table 3*) as described above for carrageenan may be employed to entrap plant cells in agar or agarose. No treatment with potassium is, however, required with these polymers.

2.1.4 *Polyacrylamide*

Polyacrylamide has been extensively used for the immobilisation of microbial cells. The chemicals used in the radical polymerisation of acrylamide are, however, relatively toxic to plant cells and this immobilisation method has therefore not been used in our laboratory to any great extent.

However, methods have recently been developed to reduce the toxicity of acrylamide and other reagents (11,12) and plant cells have been immobilised in polyacrylamide gels with preserved viability (13,14).

An elegant way to eliminate the contact between cells and toxic compounds has been reported by Freeman and co-workers (11,13). Initially a linear acrylamide polymer containing hydrazine groups is prepared and, after mixing with the cells,

Table 3. Methods for the Entrapment of Cells in Carrageenan.

Procedure A[a]

1. Dissolve low-gelling temperature carrageenan (3% w/v; FMC, NJAL 798) in 0.9% NaCl by heating and subsequently autoclaving (20 min at 121°C).
2. Suspend the collected plant cells (2.0 g wet weight) in carrageenan (8.0 g) at 35°C in a sterile beaker and quickly transfer them to a plastic syringe.
3. Allow the suspension to drip into a medium containing 0.3 M KCl.
4. Leave the carrageenan beads formed for 30 min in this medium and collect by filtration.
5. Wash and transfer the beads to an appropriate medium.

Procedure B[b]

1. Pour the carrageenan/cell suspension prepared as in Procedure A over a moulding form consisting of two Teflon plates (1 − 2 mm thick), one of which is tightly covered with holes (1 − 2 mm in diameter) (9). The plates are held together by clamps.
2. Spread the suspension carefully with a sterile spatula and, after cooling, remove the solidified cylindrical beads and place them in a medium containing 0.3 M KCl for 30 min.
3. Wash the beads and place them in an appropriate medium.

Procedure C[c]

1. Prepare the carrageenan/cell suspension in the same manner as in Procedure A and pour it under constant magnetic stirring into autoclaved (20 min at 121°C) vegetable oil (40 ml) maintained at 35°C.
2. When droplets of appropriate size have formed, cool the mixture on an ice-bath to around 10°C whereupon the polymer will solidify and beads with entrapped cells are obtained.
3. Transfer the mixture to centrifuge tubes containing 0.3 M KCl (one-third of the tube filled) and spin the beads down (2 min; 100 g).
4. Remove the oil phase and most of the aqueous phase using an aspirator (with a trap to collect the oil), and repeat the washing procedure until no oil is present.
5. Leave the beads in 0.3 M KCl for 30 min and then collect them by filtration.

[a]Normally, plant cell media contain a sufficient concentration of potassium to keep the carrageenan beads intact. However, it may be difficult to obtain uniform beads with this technique.
[b]This method of immobilisation can only be used to prepare small quantities of entrapped cells.
[c]Beads of uniform shape and size may be made by this procedure (10). If necessary, the beads are fractionated by sieving on sterile metal screens. The size of the beads can be adjusted by regulating the stirring speed during preparation. Carrageenan beads prepared by this procedure are shown in *Figure 3A*.

these polymer chains are cross-linked by a dialdehyde (e.g., glyoxal). *Mentha* cells immobilised by this method were viable and could carry out bioconversions [e.g. (−)menthone to (+)neomenthol].

A different approach to lower the toxicity of acrylamide was taken by Rosevear and co-workers (12,14). The cells were mixed prior to immobilisation with a viscous solution (alginate or xathan gum) and subsequently the monomer solution was added. Polymerisation was rapidly carried out and the entrapped plant cells (e.g., *Catharanthus roseus*) remained viable after immobilisation.

2.1.5 *Other Matrices*

Plant cells have been immobilised in other matrices such as polyurethane foam (15) and urethane prepolymers (16).

Cells of *Capsicum frutescens* and *Daucus carota* were immobilised by adding pre-made cubes (10 x 10 x 10 mm) of porous polyurethane to suspension cultures of the cells (15). The cells invaded, and were strongly retained in, the foam particles over a 21 day culture period. The viability of the immobilised cells was high. After inoculation into fresh medium, at least 95% of the cells remain immobilised after 3 – 4 days.

Urethane prepolymers possessing terminal isocyanate functional groups are convenient to use for the immobilisation of biocatalysts (17). When such prepolymers are mixed with an aqueous solution (cell suspension), a gel is formed within a few minutes. In the presence of water the prepolymers react with each other forming urea linkages. Cells of *Lavandula vera* were immobilised by this method and used for pigment production (15).

2.1.6 *Membrane Entrapment*

Various membrane configurations may also be used for the entrapment of plant cells. Hollow-fibre reactors were the first examples of this type of plant cell immobilisation (18,19). The plant cells are introduced on the shell side of the reactor and oxygenated medium is supplied at a relatively high flow-rate through the fibre-lumen. In a model study it was shown that cells of *Glycine max* entrapped in a hollow-fibre reactor produced phenolic compounds at a steady rate for 700 h (19). Furthermore, it has been shown that the pore size of the membrane used is of great importance for cell metabolism (20).

2.2 Adsorption Techniques

Immobilisation by adsorption to a solid support offers a very mild method of immobilisation. The technique has, however, not yet been used for the immobilisation of plant cells. The immobilisation of cells in polyurethane foam particles as described above may be due at least partly to adsorption.

Plant protoplasts have been immobilised by adsorption to microcarriers (21). Cross-linked dextran substituted with DEAE-groups (Cytodex 1, Pharmacia, Sweden) was used to adsorb plant protoplasts with preserved viability.

2.3 Covalent Linkage Techniques

Covalent coupling of microbial cells to a solid support has been used to a limited extent. There is also one example of the covalent linkage of plant cells (*Solanum aviculare*) to a solid matrix (polyphenyleneoxide) (22). The matrix was activated with glutaraldehyde and, after removal of excessive aldehyde, the gel was added to a suspension of the plant cells. The immobilised cells were used in a column reactor for the continuous production of steroid glycosides for 11 days.

3. VIABILITY OF IMMOBILISED PLANT CELLS

In most cases a viable cell preparation is required after immobilisation. This is particularly true when the whole metabolism of the cells is employed for *de novo* synthesis.

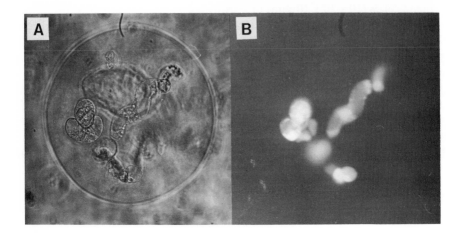

Figure 3. **A.** Carrageenan bead containing 10% (w/w) *D. carota* cells. The bead was made according to *Table 3*, procedure C. **B.** The same bead after staining with FDA.

3.1 Staining Techniques

3.1.1. *Fluorescein Diacetate*

Fluorescein diacetate (FDA) staining is commonly employed as a viability assay. The FDA is adsorbed by the cells and esterase activity within the cells will deacetylate the dye which then becomes fluorescent. The stained cells are inspected under a microscope with u.v. attachment. *Figure 3B* shows some immobilised cells stained with FDA.

3.1.2 *Carbol-fuchsin*

The mitotic index (MI) represents the percentage of the total cell population of a culture that, at a given time, exhibits some stage of mitosis. An indication of the viability may be obtained by following the MI during incubation. Carbol-fuchsin is a convenient dye to use for staining chromosomes after fixation. MI curves for freely suspended and immobilised plant cells are shown in *Figure 4*.

Details of the methodology involved in the viability assays using these stains are given in *Table 4*.

3.2 Respiration

Respiration of cells can be used as an indication of cell viability. This is particularly true when the respiration is measured at various times during incubation. Respiration is monitored with an oxygen electrode of the Clark type in the following standard procedure.

Incubate free cells (100 – 200 mg fresh weight) or the corresponding amount of immobilised cells in medium (total volume 5 ml) at 25°C and follow the oxygen consumption on a recorder. Determine the dry weight of the cells (see below) and calculate a specific respiration rate.

Immobilised cells often show a lower respiration than the corresponding

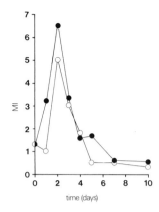

time (days)

Figure 4. Mitotic index for freely suspended (open symbols) and agarose-entrapped (solid symbols) cells of *C. roseus*.

Table 4. Staining Techniques for Measuring the Viability of Immobilised Plant Cells.

Fluorescein diacetate

1. Prepare a stock solution (0.5%) of FDA in acetone and store it at $-20°C$. A staining solution is prepared fresh daily by adding the stock solution dropwise to 10 ml medium until the first persistent milkiness appears.
2. Add this solution (1:1) to suspended or immobilised cells and after 5 min inspect the mixture under the fluorescence microscope (excitation filter $330-380$ nm; emission filter >420 nm).
3. Determine the total number of cells as well as the number of fluorescent cells and calculate the percentage of viable cells.
4. Cut the immobilised cell preparation into thin slices with a scalpel before the analysis.

Carbol-fuchsin

1. Treat free or immobilised cells with fixative [glacial acetic acid:ethanol (1:3)] for 30 min. N.B. The immobilised cell preparation should be cut into thin slices before the fixation.
2. Transfer a sample of fixed cells to a microscope slide and add an equivalent amount of carbol-fuchsin solution (3 g in 100 ml 70% ethanol).
3. After a few minutes cover the sample with a cover slip and heat until small bubbles appear.
4. Cover the sample with a paper tissue, press gently and subsequently inspect it in the microscope and count the red stained nuclei. Count at least 500 nuclei and determine the number of nuclei in mitosis. The mitotic index is calculated:

$$MI = [\text{number of nuclei in mitosis/total number of nuclei observed}] \times 100.$$

amount of freely suspended cells as a result of diffusion barriers within the polymeric network, as illustrated in *Figure 5*.

3.3 Cell Growth and Division

Increase in cell mass and cell number are good measurements of cell viability. The cell number is, however, often very difficult to determine for a plant cell culture as the cells grow in aggregates of various sizes. The cell mass (fresh weight or dry weight), is, on the other hand, relatively simple to determine. The dry weight of immobilised cells can be determined assuming the dry weight of the matrix stays

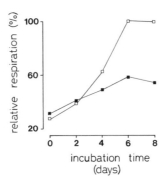

Figure 5. Respiration of alginate-entrapped cells of *C. roseus* (solid symbols) and of the same preparation after re-dissolving the polymer (open symbols).

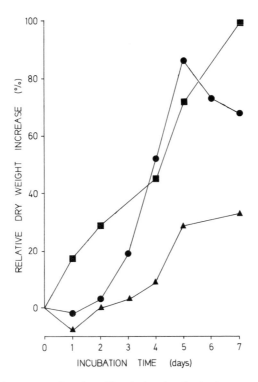

Figure 6. Dry weight increase as a function of incubation time for freely suspended (— ● —), agarose-entrapped (— ■ —), and alginate-entrapped (— ▲ —) cells of *Glycine max* (soy bean).

constant during incubation. In *Figure 6* the dry weight increase of some plant cell preparations as a function of incubation time is shown. This may be determined as follows:

(i)   Inoculate a sufficient number of 25 ml Erlenmeyer flasks containing 10 ml of medium with free cells (0.5 – 1.0 g fresh weight) or the corresponding amount of immobilised cells.

(ii)   Take samples each day or every second day and determine the dry weight by collecting the cells on pre-weighed dry filter papers.

(iii) Wash the cells on the filter with water and subsequently dry at 60 – 80°C until a constant weight is observed.

(iv)   For the immobilised cell preparation, subtract the dry weight of the polymer.

Cell division may be determined according to the method described above for calculation of the mitotic index.

4. BIOSYNTHETIC CAPACITY OF IMMOBILISED PLANT CELLS

The broad spectrum of chemical reactions taking place in cells from higher plants may be employed for the synthesis of very complex compounds. These reactions range from bioconversions (e.g., hydroxylations, methylations, etc.) to *de novo* synthesis from a simple carbon source. Immobilised viable plant cells appear to have the same biosynthetic capacity as freely suspended cells. Some representative examples of the biosynthetic capacity of immobilised plant cells are given below.

4.1 **Bioconversions**

4.1.1 *12-β-Hydroxylation*

The 12-β-hydroxylation of a digitoxin derivative to the corresponding digoxin derivative is of commercial interest. Cultured cells of *Digitalis lanata* can carry out this reaction in free suspension or in the immobilised state in batch (23) or continuous operation (24) as described in *Table 5*.

The same preparation of alginate-entrapped *Digitalis* cells as described in *Table 5* can be utilised in a batch operation (by transferring the beads to fresh medium every third day) for 180 days for the hydroxylation of methyldigitoxin to methyl-digoxin (25). This utilisation time is considerably longer than that observed for

---

Table 5. 12-β-Hydroxylation of Digitoxin by Immobilised Cells of *D. lanata*.

Method 1.

1.   Suspend cells of *D. lanata* (3 g wet weight) in 3% alginate (10 g; Sigma type IV) and im-mobilise as described in *Table 1*.
2.   Incubate the immobilised cells (30 beads) in medium (25 ml) containing digitoxin (500 μg) on a shaker at 100 r.p.m.
3.   Analyse samples of the medium for digoxin content by radioimmunoassay. The cells produce digoxin for at least 33 days.

Method 2.

1.   Immobilise *D. lanata* cells having an increased hydroxylation power in alginate and pour them into a column (3 x 20 cm).
2.   Pump a 50 μM digitoxin solution through the column at a rate of 22 ml/h.
3.   Determine the content of digoxin in the effluent by radioimmunoassay. The maximum yield is 77% and after 70 days continuous operation the column shows between 70 and 80% of this activity.

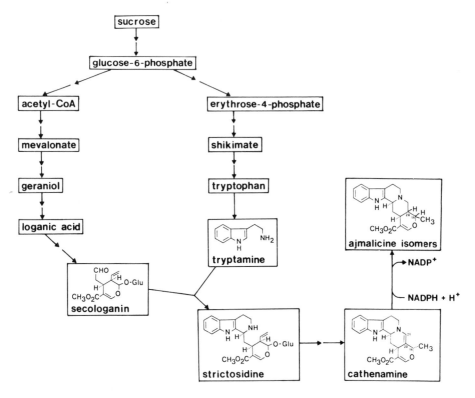

Figure 7. Biosynthetic pathway for the synthesis of ajmalicine isomers by cells of *C. roseus*. Two arrows between substances indicate two or more enzymes. The isomers formed are ajmalicine (19-βH; 20-β-H), 19-epi-ajmalicine (19-αH; 20-βH), and tetrahydroalstonine (19-βH; 20-αH).

freely suspended cells. Consequently, a higher product yield per biomass unit is obtained with immobilised cells.

4.1.2 *Reduction of Double Bond*

Cultured cells of *Catharanthus roseus* produce various indole alkaloids, such as ajmalicine, serpentine and catharanthine, depending on the cell line. The last enzyme of the biosynthetic pathway leading to ajmalicine is cathenamine reductase (see *Figure 7*). Agarose-entrapped permeabilised cells of *C. roseus* were employed for the conversion of cathenamine to ajmalicine isomers (26). The relative yields of product obtained in various incubation mixtures are summarised in *Table 6*. The methodology for this bioconversion is given in *Table 7*.

4.2 **Synthesis from Precursors**

When the biosynthetic pathway for a product is known and when appropriate precursors are available the yield of product can be increased considerably by addition of these precursors. The biosynthetic pathway for ajmalicine isomer synthesis by cells of *C. roseus* has been extensively studied (27) and tryptamine and secologanin are two distant precursors, as illustrated in *Figure 7*. The synthesis of

Table 6. Conversion of Cathenamine to Ajmalicine Isomers by Agarose-entrapped Cells of *C. roseus.*

Assay mixture	Relative alkaloid formation (%)
Complete[a]	100
$-$ NADPH, $+$ NADP$^+$ and isocitrate	85
$-$ NADPH, $+$ isocitrate	35
$-$ NADPH	25
$-$ cathenamine	9

[a] 1 g beads in 10 ml 50 mM potassium phosphate buffer, pH 7.5, containing 11 μM cathenamine, 1 mM NADPH and 28 mM mercaptoethanol.

Table 7. Bioconversion of Cathenamine to Ajmalicine by Immobilised Cells of *C. roseus.*

1. Mix dimethylsulphoxide (DMSO) treated (10% DMSO for 30 min) cells of *C. roseus* (5 g wet weight) with 5% agarose (5 g; Sigma type VII) at 35°C.
2. Make beads according to the methods described in *Table 3* Procedure C.
3. Incubate the beads (1 g) in 10 ml of 50 mM potassium phosphate buffer, pH 7.5, containing 28 mM mercaptoethanol in addition to various combinations of substrates, as summarised in *Table 6.*
4. Follow the conversion of cathenamine to ajmalicine isomers by removing samples from the incubation mixture for analysis by reversed-phase h.p.l.c.

Table 8. Synthesis of Ajmalicine by Immobilised Cells of *C. roseus.*

Method 1.

1. Mix cells of *C. roseus* (10 g wet weight) with 2% alginate (10 g; Sigma type IV) and make into beads as described in *Table 1.*
2. Pack the beads in a column.
3. Recirculate medium (10 ml) containing tryptamine [25 μmol; labelled with ^{14}C (233 000 d.p.m./μmol)] and secologanin (25 μmol) through the column at a flow-rate of 25 ml/h.
4. Continuously extract lipophilic products from the aqueous phase by bubbling the effluent through chloroform (10 ml) before it is recirculated according to the flow diagram shown in *Figure 8.*

Method 2.

1. Immobilise cells of *C. roseus* in alginate, agarose, agar and carrageenan according to the methods previously described, giving a final cell concentration of 50% (w/w) for all four preparations.
2. Add the immobilised cells (1.0 g beads) or the corresponding amount of free cells (control) to medium (10 ml) containing tryptamine [1.25μmol; labelled with ^{14}C (400 000 d.p.m./μmol)] and secologanin (1.25 μmol) and incubate on a shaker at 100 r.p.m.
3. Analyse samples for ajmalicine content after 2 and 5 days. The results are summarised in *Table 9.*

ajmalicine from these precursors by immobilised cells of *C. roseus* has also been investigated (9,23) and the methodology is shown in *Table 8.*

4.3 De novo Synthesis

De novo synthesis of complex secondary products from a simple carbon source by cultivated plant cells involves large parts of the cell metabolism. Unless the cells are viable such synthesis cannot be expected. There are quite a few examples

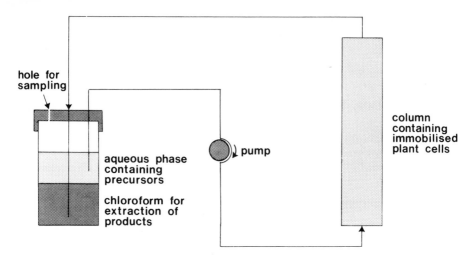

Figure 8. Schematic diagram of recirculated column reactor with continuous extraction of lipophilic products.

Table 9. Relative Production of Ajmalicine from Tryptamine and Secologanin by Free and Immobilised Cells of *C. roseus*.

Cell preparation	Relative product yield (%)	
	2 days	*5 days*
Free cells	91	100
Alginate-entrapped cells	125	176
Agarose-entrapped cells	99	114
Agar-entrapped cells	81	95
Carrageenan-entrapped cells	70	82

on the *de novo* synthesis of secondary products by immobilised plant cells and it is clear that the immobilisation *per se* does not change the biosynthetic capacity of the cells to any great extent. Here two examples will be described, i.e. the *de novo* synthesis of indole alkaloids and anthraquinines.

4.3.1 *Indole Alkaloids*

Indole alkaloids, such as ajmalicine and serpentine, have been produced by utilising alginate-entrapped (9,24) and polyacrylamide-entrapped (13) cells of *C. roseus*. The cell lines used have been selected for the production of these alkaloids. The biosynthetic pathway of ajmalicine is shown in *Figure 7*.

(i) Immobilise cells of *C. roseus* in alginate, agarose, agar and carrageenan as previously described.

(ii) Incubate the immobilised cells (1.0 g beads) or the corresponding amount of freely suspended cells (control) in a hormone-free medium (10 ml) on a shaker at 100 r.p.m. This medium is used to prevent extensive growth of cells which could lead to leakage of cells from the beads.

Table 10. Relative Production of Ajmalicine by *de novo* Synthesis from Sucrose by Free and Immobilised Cells of *C. roseus* after 2 Weeks Incubation.

Cell preparation	Relative product yield (%)
Free cells	100
Alginate-entrapped cells	140
Agarose-entrapped cells	100
Agar-entrapped cells	88
Carrageenan-entrapped cells	62

(iii)   After 2 weeks take samples and analyse them for ajmalicine content by reversed-phase h.p.l.c.

The results are summarised in *Table 10*.

4.3.2 *Anthraquinones*

Anthraquinones are produced by various plant cell cultures. The production of anthraquinones by alginate-entrapped cells of *Morinda citrifolia* has been investigated (28). In these studies it was observed that the immobilised cells produced around 10 times as much of the product as freely suspended cells under the same conditions.

(i)   Mix cells of *M. citrifolia* (15 g wet weight) with 3% alginate (25 ml; Sigma type IV) and 5 ml medium whereupon beads are made.

(ii)   Incubate the immobilised plant cells (30 beads) in hormone-free medium (25 ml) on a shaker at 100 r.p.m. at 23°C.

(iii)   During cultivation take samples and analyse them for anthraquinone content as well as for the number of cells within the beads.

(iv)   Collect the beads from one flask and wash them.

(v)   Extract one portion of the beads (20) with 80% ethanol for 30 min and determine the amount of extracted anthraquinone spectrophotometrically at 434 nm.

The remaining beads can be dissolved in 1 M potassium phosphate buffer, pH 6.5 (1.0 ml) to determine the number of cells in a haemocytometer. The amount of anthraquinone after 22 days is 1 and 9.5 pmol/cell for freely suspended and immobilised cells, respectively.

5. PRODUCT RELEASE

Secondary products are often stored within cultured cells (in the vacuoles) and, to a certain extent, this limits the applicability of immobilised plant cells. One of the major advantages of an immobilised biocatalyst is the possibile continuous operation of a process. This requires the release of product into the surrounding medium (extracellular products) and therefore attempts are now being made to induce release of intracellularly stored products from the immobilised cells. The immobilisation as such can induce such a product release (spontaneous release) but in most cases it appears to be necessary to induce product release by additions to the medium (permeabilisation).

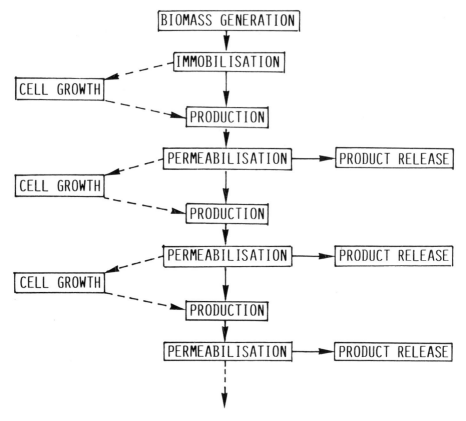

Figure 9. Schematic diagram of a process for the production of intracellular plant biochemicals using intermittent product release from immobilised plant cells. The growth phase is optional.

5.1 Extracellular Products

Some plant products are extracellular and an immobilised preparation can be used in a continuous process without difficulties. For instance, as described above, cells of *D. lanata* were used in a continuous column operation for 70 days to convert digitoxin to digoxin.

5.2 Intracellular Products

5.2.1 *Spontaneous Product Release*

In a few cases it has been observed that products, which are normally stored within cultured cells, are released into the medium after immobilisation (24,29). In this way indole alkaloids were excreted from immobilised cells of *C. roseus*. The mechanism behind this spontaneous release is not fully known but it has been suggested that the proximity of cells within a gel particle is causing the release of product (29).

5.2.2 *Induced Product Release*

The spontaneous release described above is not of a general character and therefore methods are being worked out by which immobilised cells can be induced to excrete the desired product (26,30). It has been shown that it is possible to release products by permeabilisation of the cell membranes with DMSO without affecting the viability or biosynthetic capacity of the immobilised cells (30). This means that the same preparation of immobilised cells may be utilised for the production of many batches of product in a cyclic manner including a permeabilisation step, as schematically shown in *Figure 9*.

(i)   Immobilise cells of *C. roseus* in agarose at a final cell concentration of 20% (w/w).
(ii)   Incubate the immobilised cells (2.0 g beads) batch-wise in hormone-free medium (10 ml) containing 0.5 mM ^{14}C-labelled tryptamine (350 000 d.p.m./μmol) and 0.5 mM secologanin for 3 days at 26°C.
(iii)   Collect the beads and wash them with medium (2 x 10 ml) and place them in medium containing 5% DMSO for 30 min.
(iv)   Wash the DMSO-treated beads with medium (2 x 10 ml) and then transfer them to complete medium (growth medium) and incubate for 2 – 4 days.
(iv)   Repeat the production/growth cycle.
(vi)   Combine the media collected and extract with methylenechloride (3 x 10 ml) at pH 10.
(vii)   Analyse the extracts for radiolabelled tryptamine by t.l.c. (9).

The result of a typical experiment carried out according to the procedure described above (see *Figure 10*) shows that it is possible to release products stored within immobilised plant cells by a permeabilisation procedure without affecing cell metabolism. A continuous process based on intermittent release of product can be designed for the production of intracellular plant products.

6. CONCLUDING REMARKS

The first report on the immobilisation of plant cells in a beaded polymer appeared in 1979 (23) and since then an increasing number of reports have occurred. It is believed that by immobilising plant cells some major problems in utilising this type of cell on a large scale may be reduced or even overcome

After immobilisation the cells may grow to a certain extent before they start to leak into the surrounding medium but it would not be advisable to produce compounds which are directly associated with cell growth when using entrapped cells. It is therefore likely that immobilised plant cells only find applications for products that are non-growth associated, i.e., they are produced by the cells in stationary phase. The extended stationary phase observed for immobilised plant cells may circumvent some of the problems resulting from the apparent slow growth of cells (biomass generation).

Cell aggregation occurs in culture and this leads to a wide distribution in aggregate size resulting in problems with agitation of, and mass transfer to, the cells. These types of problems may be reduced by immobilisation since 'aggregation' becomes a design parameter. A relatively homogeneous preparation of

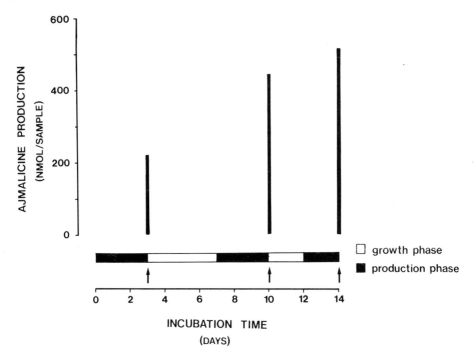

Figure 10. Ajmalicine production from precursors by agarose-entrapped cells of *C. roseus* according to the scheme shown in *Figure 9*. The cells were intermittantly permeabilised (arrows) for complete product release.

beads containing the cell aggregates is readily made.

Plant cell cultures often show significant changes upon long-term serial passages. For instance, cell lines selected for high productivity of a particular compound frequently lose this capability. The extended stationary phase observed with immobilised cells may also, to a limited extent, reduce this problem since the cells are utilised in a non-dividing state which should reduce the appearance of genetic changes.

The problem of low shear tolerance of plant cells is eliminated by the entrapment of the cells in a protective polymeric matrix. This results in a simpler reactor design.

The possibility of releasing intracellularly stored products by intermittent permeabilisation of the immobilised cells is yet another advantage that may prove valuable in the future development of plant tissue culture technology. This allows the re-utilisation of the biomass over an extended period of time.

The results so far obtained are promising and immobilisation of plant cells may prove a valuable tool in overcoming or reducing some of the problems encountered in the utilisation of plant tissue cultures for the production of complex natural products.

7. REFERENCES

1. Zenk,M.H. (1978) in *Frontiers of Plant Tissue Culture*, Thorpe,T.A. (ed.), International Association for Plant Tissue culture, p. 1.
2. Staba,E.J., ed. (1980) *Plant Tissue Culture as a Source of Biochemicals*, published by CRC Press, Boca Raton, FL.
3. Yamada,Y. and Fujita,Y. (1983) in *Handbook of Plant Cell Culture*, Vol. 1, Evans,D.A., Sharp, W.R., Ammirato,P.V. and Yamada,Y. (eds.), Macmillan Publishing Co., New York, p. 717.
4. Brodelius,P. (1978) in *Advances in Biochemical Engineering*, Vol. 10, Ghose,T.K., Fiechter,A. and Blakebrough,N. (eds.), Springer-Verlag, Berlin, Heidelberg, New York, p. 76.
5. Mosbach,K., ed. (1976) *Methods in Enzymology,* Vol. 46, published by Academic Press, New York.
6. Brodelius,P. and Mosbach,K. (1982) in *Advances in Applied Microbiology*, Vol. 28, Laskin,A.I. (ed.), Academic Press, New York, p. 1.
7. Brodelius,P. (1983) in *Immobilized Cells and Organelles,* Vol. 1, Matthiasson,B. (ed.), CRC Press, Boca Raton, FL, p. 27.
8. Brodelius,P. (1983) *Ann. N.Y. Acad. Sci.,* **413**, 383.
9. Brodelius,P. and Nilsson,K. (1980) *FEBS Lett.,* **122**, 312.
10. Nilsson,K., Birnbaum,S., Flygare,S., Linse,L., Schroder,U., Jeppsson,U., Larsson,P.-O., Mosbach,K. and Brodelius,P. (1983) *Eur. J. Appl. Microbiol. Biotechnol.,* **17**, 319.
11. Freeman,A. and Aharonowitz,Y. (1981) *Biotechnol. Bioeng.,* **23**, 2747.
12. Rosevear,A. (1981) *European Patent Application*, 81304001.1.
13. Galun,E., Aviv,D., Dantes,A. and Freeman,A. (1983) *Planta Med.,* **49**, 9.
14. Lambe,C.A. and Rosevear,A. (1983) *Proceedings of Biotech 83,* London, May 4−6, 1983, p. 565.
15. Lindsey,K., Yeoman,M.M., Black,G.M. and Mavituna,F. (1983) *FEBS Lett.,* **155**, 143.
16. Tanaka,A., Sonomoto,K. and Fukui,S. (1983) Poster I:15 presented at *7th International Conference on Enzyme Engineering,* September 25−30, 1983, White Haven, PA.
17. Fukui,S. and Tanaka,A. (1984) in *Advances in Biochemical Engineering,* Vol. 29, Fiechter,A. (ed.), Springer-Verlag, Berlin, Heidelberg, New York, p. 1.
18. Shuler,M. (1981) *Ann. N.Y. Acad. Sci.,* **369**, 65.
19. Jose,W., Pedersen,H. and Chin,C.K. (1983) *Ann. N.Y. Acad. Sci.,* **413**, 409.
20. Shuler,M. and Hallsby,A.G. (1983) Paper #76b presented at *AIChE Summer National Meeting,* Denver, Colorado, August 28−31, 1983.
21. Bornman,C.H. and Zachrisson,A. (1982) *Plant Cell Rep.* **1**, 151.
22. Jirku,V., Macek,T., Vanek,T., Krumphanzl,V. and Kubanek,V. (1981) *Biotechnol. Lett.,* **3**, 447.
23. Brodelius,P., Deus,B., Mosbach,K. and Zenk,M.H. (1979) *FEBS Lett.,* **103**, 93.
24. Brodelius,P., Deus,B., Mosbach,K. and Zenk,M.H. (1980) *European Patent Application* 80850105.0.
25. Alfermann,A.W., Bergmann,W., Figur,C., Helmbold,U., Schwantag,D., Shuller,I. and Reinhard,E. (1983) in *Plant Biotechnology*, Mantell,S.H. and Smith,H. (eds.), Cambridge University Press, Cambridge, p. 67.
26. Felix,H., Brodelius,P. and Mosbach,K. (1981) *Anal. Biochem.,* **116**, 462.
27. Zenk,M.H. (1980) *J. Nat. Prod.,* **43**, 438.
28. Brodelius,P., Deus,B., Mosbach,K. and Zenk,M.H. (1980) in *Enzyme Engineering*, Vol. 5, Weetall,H.H. and Royer,G.P. (eds.), Plenum Press, New York, p. 373.
29. Rosevear,A. and Lambe,C.A. (1982) *European Patent Application* 82301571.4.
30. Brodelius,P. and Nilsson,K. (1983) *Eur. J. Appl. Microbiol. Biotechnol.,* **17**, 275.

CHAPTER 9

Immobilised Mammalian Cells in Hormone Detection and Quantitation

STEPHEN P.BIDEY

1. INTRODUCTION

1.1 The Immobilised Mammalian Cell

Recent years have witnessed major advances in the development of techniques associated with the controlled maintenance, under *in vitro* conditions, of cells derived from mammalian tissues. Several factors may be identified which have directly contributed to the rapid progress in this area of biotechnology, including the isolation of continuous, fully-characterised cell lines, the identification of specific cellular growth factors and the development of cell culture apparatus designed to maintain isolated cells under optimal *in vitro* conditions. The combined outcome of these individual contributions has been that virtually any mammalian cell type may now be maintained in culture in a state of high functional differentiation. This rapid evolution in cell culture metholodgy has been of enormous benefit to research in many areas of the biological sciences, whilst the isolation and study of cellular products such as hormones, enzymes and antibodies has been progressively facilitated.

Depending upon the requirement of isolated cells for attachment to a solid support or matrix, two distinct techniques of cell culture have evolved. Most cells derived from solid tissue such as kidney, liver and skin are anchorage-dependent, and may only be maintained under *in vitro* conditions as a sheet or 'monolayer' in close association with a supporting surface. The individual cells from which such preparations are formed establish contacts both with adjacent cells and with the support itself and may thus be considered as immobilised. However, in contrast to the immobilisation of enzymes, antibodies and microbial cells, which frequently require chemical modification in order to effect the coupling procedure, the bonding between anchorage-dependent mammalian cells and a solid support is typically attainable without deliberate physical or chemical intervention, given the initial viability of the isolated cell preparation, compatability with the solid support and the presence of attachment-promoting factors either in serum or as a defined supplement to the culture medium.

A second type of cell culture system, the stirred fermentor culture, has been developed to meet the specific requirements of certain cell types (e.g., those of lymphoblastoid origin), which may be maintained under conditions analogous to those used to support the proliferation of bacterial cells. Thus, the ability of such cells to grow in suspension, without prior immobilisation, has facilitated the development of industrial-scale cell propagation techniques. Until relatively recent times, the large-scale culture of anchorage-dependent cells was limited by the excessive surface area required for immobilisation, and an adequate but precisely-controlled supply of nutrients and oxygen. Although recent developments in microcarrier culture techniques, using cells maintained as immobilised monolayers on polystyrene, glass or dextran beads, have enabled certain anchorage-dependent cell types to be maintained under suspension-culture conditions, problems associated with the control of culture environment have yet to be overcome before the full potential of this technique as an industrial-scale process can be realised.

In many other areas of the biological sciences, and particularly in the field of clinical research, immobilised anchorage-dependent cells are playing an increasingly important role, having contributed significantly to our present understanding of cell growth regulation and differentiation, cell ageing, and the functional and morphological differences between normal and tumour cells. The high degree of functional differentiation typically retained by immobilised cells has also facilitated studies of the control of cellular metabolism by specific hormones and defined growth factors, and modification of cellular function by drugs and toxins. In this respect, widespread use has been made of immobilised hepatocytes in studies of the metabolism of foreign compounds by the liver. Whilst many of these studies have been of a semi-qualitative nature, the recent rapid advance in cell culture biotechnology has enabled immobilised cell monolayers to be used as the basis of sensitive, specific and highly quantitative detection systems, collectively referred to as bioassays, for a range of molecules of biological significance. In particular, the recent application of such *in vitro* diagnostic techniques to clinical and experimental endocrinology has facilitated the determination of circulating hormone levels in both healthy and diseased individuals, and has led to a proliferation of research activities in the areas of hormone biosynthesis, metabolism and cellular responses.

1.2 The Bioassay — Some Basic Concepts

Fundamentally, two distinct types of hormone detection and quantification system may be identified, based upon structural and functional properties of the hormone molecule, respectively. Those techniques involving the detection of, and discrimination between, discrete molecular forms of a hormone have been classified as 'structurally-specific' (1) and rely upon the specificity of naturally-occurring binding proteins, or experimentally-raised antibodies for the unique molecular conformation of any given hormone. Such systems, which include radioimmunoassay as the best known and most widespread example, are designed to quantify the amount (i.e., the number of molecules) of a given structure in a test sample relative to standard preparations containing known amounts of the same material, by analysis of hormone-antibody interactions. Significantly, however, these 'structurally-specific' systems take no account of the biological efficacy or 'potency' of the material under investigation and hence are unable to differentiate between biologically-active and inactive forms of an individual hormone.

In contrast to the essentially structural quantitations and comparisons underlying radioimmunoassay and related methodologies, bioassay techniques may be classified as 'functionally-specific' since the underlying principle employed is a comparison of the *effects* of different amounts of substances (e.g., hormones and enzymes) sharing a common physiological function. In practice, such systems essentially compare the magnitudes of response induced in a target tissue after incubation with test and control stimulators, respectively. The latter should be identical with regard to their specificity for the particular target tissue used. The underlying concepts of bioassay are illustrated in *Figure 1*. It is particularly important to recognise that the magnitude of the specific cellular response induced by the stimulator will be proportional to the degree of occupancy of cell-surface receptors, as a direct function of the 'concentration' of the biologically active molecule in the test sample to which the target tissue is exposed. The 'functional specificity' of the target tissue response will ensure that only those molecules having intrinsic bioactivity with respect to the target tissue will be 'seen' by that tissue, whilst other molecules, even those having close structural similarities to the native stimulator, will not invoke a response, and will thus not be 'measured'. However, closely analogous molecular forms may be indistinguishable from the biologically 'intact' molecule when investigated using 'structurally-specific' methods.

The gradual development of techniques whereby appropriate 'target' tissues for a range of molecules of biological interest may be isolated and maintained *in vitro* has enabled the development of specific *in vitro* systems to determine the

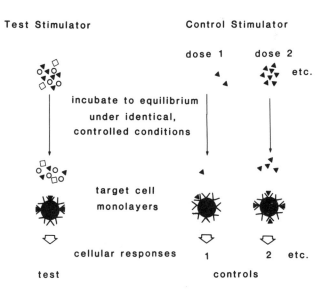

Figure 1. Bioassay for a specific stimulator in a heterologous ligand population, based upon a defined functional response of cells immobilised as target 'tissue' in monolayer culture.

biological activity or 'potency' of hormones in serum and other biological fluids. This has allowed the simultaneous assay for certain hormones using both 'structurally-specific' and 'functionally-specific' techniques, which in certain disease states has revealed discrepancies between hormonal bioactivity and mass (i.e., specific activity), relative to that found in normal individuals. Clearly therefore, a need exists for both types of assay system in investigative clinical medicine.

It is hoped that this chapter will fulfil three principal objectives. In the first instance, it is intended to provide the reader with a basic insight into the fundamental criteria and practical considerations associated with the design and construction of bioassay systems for hormones, based upon functional responses of cells immobilised as monolayers on solid surfaces. Secondly, it is intended to serve as a practical guide to the procedures currently in use in the author's laboratory, relating to cellular isolation and immobilisation, together with the setting up and running of a simple bioassay based upon immobilised primary cultures of cells derived from whole tissue. Finally, the importance of maintaining appropriate and adequate control over bioassay performance will be discussed, and simple procedures described whereby such a system of monitoring may be carried out.

2. IMMOBILISED CELLS AS BIOASSAY TARGET TISSUE

The evolution of bioassay techniques based upon the responses of cultured cells has been facilitated by the rapid development of improved cell isolation and *in vitro* maintenance procedures, reflecting recent advances in cell culture technology as a fundamental tool in cell biology. In many respects, the application of immobilised cells to the bioassay of hormones in biological fluids or tissue extracts may be regarded as a natural progression from earlier, more cumbersome and less precise procedures involving organ cultures or essentially short-term cell suspensions. In other cases, laboratory animals, hitherto used as the basis of '*in vivo*' bioassays, have been replaced directly, with a considerable gain in bioassay performance and marked simplification of bioassay design.

2.1 Supports for Cellular Immobilisation

The long-term survival, proliferation and functional stability of cells isolated by enzymatic dispersion of whole tissue are critically dependent upon adherence of the cells to a solid support or growth matrix, and it is therefore important to adopt culture procedures and immobilisation supports consistent with optimal cellular adhesion characteristics. Whilst limited success may be obtained using conventional glass culture vessels and Petri dishes, disposable, single-use polystyrene culture plates and dishes, pre-sterilised by gamma irradiation, have now virtually replaced glassware in routine cell culture applications, and the excellent optical properties of polystyrene ensure ease of visualisation of immobilised cells by phase-contrast microscopy.

All cells isolated from vertebrate tissues have a negative surface charge, and for successful immobilisation require a negatively-charged glass or polystyrene culture surface, or a polylysine-coated surface possessed of a positive charge. In

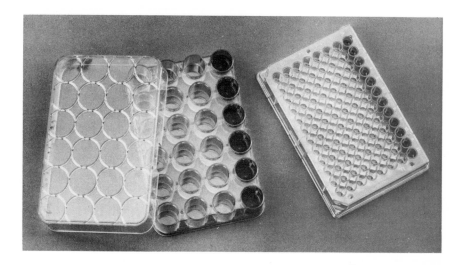

Figure 2. Polystyrene plates, used for culturing bioassay 'target' cells as immobilised monolayers. **Left**: 24-well plate, culture area 1.9 cm² per well (Linbro, Flow Laboratories Limited). **Right**: 96-well microtitre plate; culture area 0.32 cm² per well (Gibco-European Limited). Both plate types are supplied with surfaces treated for optimal cell attachment.

either case, however the charge density, rather than the polarity, is the critical factor influencing the attainment of optimal cellular attachment. Culture vessels for cellular immobilisation are therefore subjected to surface treatment during manufacture to ensure the attachment and stability of a wide variety of cell types. Two of the most widely-used types of polystyrene plates used for cell immobilisa-tion are illustrated by the 24- and 96-well flat-bottomed plates shown in *Figure 2*. However, use of the latter type is generally restricted to those bioassays in which only the incubation medium is sampled after cell stimulation, recovery of the cells being impracticable on account of the small dimensions of the wells.

2.2 The Attainment of Cellular Immobilisation

The immobilisation of anchorage-dependent cells on a solid support is a multi-stage process, and is preceded by adsorption of attachment factors on to the negatively or positively charged support surface. These factors include glycopro-teins derived either from the serum supplement, or from certain types of cells such as diploid fibroblasts, a frequent contaminant of primary epithelial-type cultures. In addition, further conditioning factors are derived from epithelial cells, and mediate attachment of the cells to the growth support by binding both to cell surface glycoproteins and to the serum or fibroblast-derived glycoproteins previously adsorbed. In this manner, the initial contact between the cell surface and the solid support leads to immobilisation of the cells, followed by a marked spreading and flattening of the cytoplasm to attain a monolayer configuration (*Figure 3*). Attainment of an intimate association between cell surface and culture support, and, by implication, the structural and functional stability of the

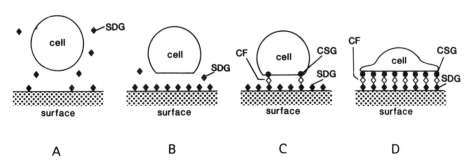

SDG : Serum–Derived Glycoproteins
CSG : Cell–Surface Glycoproteins
CF : Conditioning Factors

A   B   C   D

Figure 3. Attachment and spreading of freshly isolated cells in monolayer culture. (**A**) Glycoprotein molecules (♦) derived from the serum-supplemented culture medium are initially adsorbed onto the culture vessel surface as a result of surface-charge effects. (**B**) Cells prepared from whole tissue settle onto the surface of the culture vessel within 1 − 2 h of plating. (**C**) Conditioning factors (◇) derived from the cells mediate attachment of cells to the culture surface through binding to both serum-derived and cell-surface (●) glycoproteins. (**D**) Immobilised cells subsequently spread and flatten to assume a monolayer configuration, after approximately 24 h.

monolayer, are therefore critically dependent upon the presence of glycoprotein attachment factors in the culture medium (2).

2.3 The Cellular Environment after Immobilisation

The immobilisation of hormonally-responsive cells onto a solid support provides the investigator with ideal experimental conditions under which the various manipulations involved in a bioassay procedure may be undertaken. Thus the addition and withdrawal of test and control stimulators may be readily achieved without significant disturbance of the cell monolayer itself and with a minimal 'carry-over' of culture medium between incubation stages. The release of cellular reaction products may be readily determined by sampling of the medium at intervals during the test incubation, whilst intracellular products or metabolites may be recovered and investigated after final lysis and treatment of the post-incubated cells wtih appropriate reagents. Finally, the unique spatial configuration of cells. maintained as immobilised monolayers, which allows for the rapid and simultaneous exposure of all cells in the culture to a stimulator, is consistent with a high level of within-culture uniformity in cellular responsiveness which, in combination with the close structural identity of replicate cultures, contributes greatly to the high performance characteristic of bioassays based upon such preparations.

2.4 The Choice of Appropriate Bioassay Target Cells

The choice of target cell cultures most appropriate to the bioassay for any given hormone will reflect both the nature of the hormone under investigation, and the *in vitro* stability and response characteristics of the cells themselves. In this respect, the ultimate goal in bioassay development has been the availability of continuous cell strains or established cell lines having stable response and growth characteristics over an indefinite period, thus fulfilling the criteria required of an ideal bioassay target tissue. However, in many cases, continuous cultures have not yet been isolated, and such bioassay systems as have been developed are based upon cells isolated directly from human or animal tissues by enzymatic dispersion. Invariably, these cells will be derived from the corresponding natural target tissue for the hormone *in vivo*, and should, ideally, originate from the same animal species. However, in the case of bioassays of human cell stimulators, e.g., hormones, this requirement may be difficult to fulfil on account of the impracticability of obtaining suitable tissue specimens as starting material for cell preparations, particularly those of a non-pathological nature. Accordingly, it may frequently be necessary to resort to animal tissues in order to implement a particular bioassay technique. Such a practice should, however, only be undertaken after a systematic and thorough investigation of the specificity of interaction and dose-response characteristics of a reference preparation of the hormone with regard to human and non-human target cells, respectively.

Isolated cells, prepared directly from whole tissues by enzymatic digestion, may be immobilised and maintained in primary culture for a period of 10 – 14 days, retaining a high level of specificity and responsiveness towards appropriate stimulators, as defined by the original *in vivo* functional characteristics of the tissue. Accordingly, it has been possible to use primary cultures as the basis of bioassay techniques for several hormones, although the usefulness of such preparations is subject to the limitations ultimately imposed by a finite cell lifespan and by the inherent variability in responsiveness of individual cell preparations. Unfortunately, functionally-differentiated cells in primary culture also eventually lose many of their specific functional markers, a process which may, in many cases, be accompanied by extensive overgrowth by rapidly-proliferating, 'contaminating' cell types, particularly those of a fibroblastic nature. However, it may be possible to extend the period of target cell responsiveness through selective reduction of the rate of fibroblast proliferation. This has been successfully achieved through manipulation of the culture medium composition, involving a reduction or total elimination of serum and substitution of defined cell attachment and growth factors (3 – 6).

It is a well-established observation that the *in vivo* functional characteristics of many hormonally-responsive tissues reflect not only the age and sex, but also the health and nutritional status of the animal. Accordingly, it is not surprising that after isolation of such tissues, the primary cell monolayers subsequently prepared are characterised by a variability in responsiveness when challenged with a reference preparation of a test stimulator, which reflects the *in vivo* response characteristics of the individual starting tissues (7). Whilst each cell preparation may, in itself, constitute a valid bioassay target tissue, the frequent recourse to fresh starting tissue, made necessary by the finite lifespan and progressive decline in functional responsiveness of primary cultures, may limit the usefulness of such preparations to those situations in which the requirement for uniformity of cellular responsiveness in successive bioassays is not a critical consideration. As previously described, many of the problems associated with the design and implementation of a routine, stable and reproducible bioassay, based on primary cell cultures, may ultimately be overcome through exploitation of continuous cell strains or established cell lines. By definition, such preparations possess the capacity for unlimited proliferation and, in the case of functionally-differentiated, hormone-responsive cells, exhibit stable response characteristics with respect to a given specific stimulator, circumventing both the finite lifespan and declining responsiveness characteristic of primary cultures.

2.5 Selection of an Appropriate Cellular Response

The functional interaction between a hormone and the population of receptors specific to that hormone on appropriate target cells leads, under normal *in vivo* physiological conditions, to alterations in cell metabolism, resulting in the intracellular accumulation of metabolites, depletion of precursors and, frequently, to the release of specific cellular products. However, following the removal of such tissues from the body, and enzymatic dispersion to give isolated cells, certain of the differentiated functional characteristics of the intact tissue are frequently lost such that the immobilised cell preparations may be incapable of demonstrating the full sequence of functional responses normally associated with hormonal stimulation of the tissue. Nevertheless, the degree of differentiation retained under *in vitro* conditions is frequently sufficient to allow the investigator a choice of several specific and well-defined cellular responses upon which to design a bioassay for a given hormone.

An investigation of the various metabolic responses of target cell monolayers, and a decision as to which should be exploited as the basis of a system of measurement of the biological activity of a given stimulator, should include an assessment of each response on the basis of four criteria: (i) the specificity and (ii) the sensitivity of the response in a single cell preparation, (iii) the precision of hormone measurement attained over the desired range of hormone doses and (iv) the reproducibility of the response between different cell preparations. Furthermore, since each metabolic response will require quantification, the ease and efficiency of isolation of each metabolic product, and the characteristics of the assays [e.g., radioimmunoassays or enzyme-linked immunosorbent assay (ELISA) techniques] by which each is quantified, should be considered in formulating an objective assessment of each of the cellular responses to the hormone under investigation. An initial study should examine the *specificity* of a particular response of the target cells for a single species of stimulator, since exposure of the cells to other hormones, in particular those showing close structural analogies to that under investigation, may elicit a similar response. For example, the α-subunits of the glycoprotein hormones luteinising hormone (LH), follicle-stimulating hormone (FSH), thyroid-stimulating hormone (TSH) and human chorionic gonadotrophin (hCG) are identical and, as might be predicted, some cross-reactivity has been observed between these hormones with respect to interaction with the TSH receptor (8). In most instances however, the level of *functional* cross-reactivity, and indeed of simple 'blocking' of the receptors by structurally-related, but biologically-inactive molecules, may be minimised by

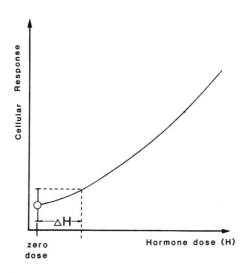

Figure 4. Calculation of bioassay sensitivity from the standard dose-response curve. $\triangle H$, the absolute error in hormone measurement at zero hormone dose is, by definition, the sensitivity of the assay, and is clearly a function of both the slope of the dose-response curve and the error in response determination (i.e., SD) at zero hormone dose.

careful initial selection of target cells and incubation conditions consistent with a high level of specific functional responsiveness with respect to the stimulator under investigation.

The second factor to be considered when assessing different responses of immobilised target cells relates to the *sensitivity* of each response. In more pragmatic terms, one should determine the smallest dose of hormone that can be detected on the basis of each of the responses being assessed. In this context, the sensitivity of a given cellular response may be defined as the lowest level of stimulator that can be distinguished from 'zero' dose under a given set of bioassay conditions, and may be estimated from the dose-response curve to the hormone under investigation, as shown in *Figure 4*. Accordingly, the relative effects on bioassay sensitivity of variations in incubation conditions, such as target cell density, exposure time and ionic strength and pH of the medium, may be assessed, and conditions consistent with optimal bioassay sensitivity established. However, in addition to estimating bioassay sensitivity, the *Precision of hormone measurement*, an index of bioassay performance, may also be estimated by constructing Precision-Dose Profiles (9). These may be used to compare bioassay performance over different dose ranges of the hormone preparation, i.e., in different regions of the standard curve, and also to investigate the influence of different incubation conditions or reagents on bioassay performance over any given range of hormone doses. In general, in optimising conditions for a given bioassay, one should seek to establish conditions consistent with the maximal precision of hormone measurement, at least within the hormone dose range of interest, and preferably over as wide a range of doses as possible. Such a formalised approach to the analysis of dose-response data was originally applied to saturation analysis data (9), and an analogous, albeit simplified, approach has subsequently been applied by the author to the optimisation and assessment of biological assay procedures for thyroid stimulators based on the *in vitro* responses of thyroid tissue slices and cell membranes (10) or whole thyroid cells maintained in primary monolayer culture (7,11).

The final consideration of importance when assessing the various functional responses of bioassay target cells relates to the *between-bioassay reproducibility* associated with each response. Using individual cell monolayers prepared from a single starting tissue, in different but successive bioassays, it is possible to attain a high degree of reproducibility, such that a between-assay variation of 10−15% should be attainable for hormone doses falling within the most precise region of the dose-response curve. A similarly uniform response should be given by successive subcultures of an established cell line. However, for reasons already discussed, such a consistently-reproducible bioassay response is frequently unattainable with primary cultures derived from different starting tissues, on account of the inherent variability in responsiveness of the latter. In such cases, a between-assay variation of the order of 40−50% is not uncommon (7).

To summarise, therefore, the response of target cell monolayers finally selected as the most appropriate functional marker for a given hormone, should be consistent with:

(i)   a high specificity of the response for the hormone under investigation;
(ii)   high sensitivity;
(iii)   good precision of measurement of the hormone over the appropriate dose-range; and
(iv)   high reproducibility in responsiveness both within and, as far as is practicable in the case of primary cell monolayers, between batches of cells prepared from different individual starting tissues.

3. PREPARATION AND IMMOBILISATION OF ISOLATED CELLS

Having introduced the reader to some of the fundamental considerations underlying the selection of target cells, immobilisation and incubation conditions, and cellular responses most appropriate to the *in vitro* bioassay of any given hormone, details will now be given of the practical procedures involved in the preparation, immobilisation and maintenance of viable, functionally-responsive cells from whole tissue, and their subsequent use as bioassay material. Throughout this section it will be assumed that the reader is familiar with basic cell culture techniques and terminology. However, for those requiring an introduction to the subject, several texts may be recommended (12,13). In the following section, a specific application of these basic procedures will be illustrated with reference to the bioassay of TSH based upon cyclic AMP (cAMP) accumulation in immobilised primary monolayers of human thyroid cells.

3.1 Isolation of Hormonally-responsive Cells from Whole Tissues

Viable and functionally-responsive cells may be isolated from either normal (e.g., biopsy specimens) or pathological human tissues, or, alternatively, from analogous tissues obtained from animals. In each case cells are isolated from

Table 1. Processing and Enzymatic Dispersion of Tissues.

Preparation of tissues

1. After removing all connective tissue from the excised tissue specimen, cut the trimmed tissue into 2 − 3 mm cubes, and transfer these to a 100 ml glass beaker.
2. Using a pair of fine-pointed curved scissors and forceps, continue to process the tissue pieces until a fine mince of fragments of 0.5 − 1.0 mm dimensions is obtained. At intervals during this procedure, wash the fragments with 20 ml portions of warm (37°C) Hank's salt solution (Ca^{2+} and Mg^{2+} free, without phenol red).
3. After washing and decanting 4 − 5 times, and with the final supernatant essentially free of erythrocytes and connective tissue debris, resuspend the washed tissue fragments in fresh Hank's salt solution, using 10 ml for each gram of original tissue processed.

Enzymatic digestion of tissue fragments

1. Transfer the washed tissue fragments, obtained after the initial processing of whole tissue specimens, into a sterile trypsinisation flask such as that manufactured by Bellco Glass Inc. (Vineland, New Jersey, USA) (*Figure 5*), and add a sterile solution of an enzymatic dispersing agent. 'Dispase II' (Boehringer-Mannheim, 1 U/ml in Hank's salt solution) has been found to be most suitable for this purpose. For each gram of tissue processed, 10 ml of the enzyme solution should be used.
2. Add a sterilised magnetic stirring bar to the flask, seal the neck of the vessel and incubate at 37°C on a magnetic stirrer at 100 − 150 r.p.m.
3. After incubation for 1 h, remove the flask from the stirrer, allow the tissue fragments to settle, and discard the supernatant, which will contain remaining erythrocytes, together with endothelial and connective tissue cells, through the side-arm of the flask.
4. Add a fresh, pre-warmed aliquot of enzyme solution to the tissue fragments and re-incubate, with stirring, for a further 1 − 2 h. At this stage in the digestion process, the cells liberated will show only a small incidence of 'contamination' with non-epithelial cells, and may be collected by centrifugation (100 *g*; 7 min) at room temperature.

Post-digestion treatment of dispersed cells

1. Resuspend the pellet of isolated cells, obtained after the enzymatic digestion of tissue fragments, in a pre-warmed (37°C) 20 ml aliquot of Hank's salt solution. Initial dispersion of the pellet may be achieved with the aid of a siliconised Pasteur pipette.
2. After obtaining a homogeneous suspension, re-centrifuge the cells (100 *g*; 7 min) to obtain a pellet.
3. Remove the supernatant, and resuspend the pellet, as described above, in 20 ml of pre-warmed Hank's salt solution.

tissue specimens by subjecting the latter to controlled enzymatic dispersion, and it is therefore important that tissues should be processed as rapidly as possible to avoid cellular necrosis and thus maximise the yield of viable cells obtained. Without exception, aseptic procedures should be adopted in all the tissue- and cell-handling manipulations described in this section.

3.1.1 *Processing and Enzymatic Digestion of Tissues*

Isolated tissue specimens obtained from either animal or human sources may be stored for short periods (not exceeding 1 h), prior to the initiation of processing, in a dry sterile specimen container at room temperature. Such conditions have been found to be consistent with a minimal loss in viability of the cell preparations subsequently isolated. Processing of the tissue prior to enzymatic dispersion

Figure 5. Trypsinisation vessel (Bellco Glassware Inc.) used to achieve enzymatic dissociation of tissue fragments to give a suspension of single cells. The side-arm allows for decantation of the cell suspension with minimal disturbance of the remaining tissue fragments.

typically involves the initial preparation of slices (0.5 – 1.0 mm thick) using, for example, a Stadie-Riggs microtome (distributed in the UK by Arnold R.Horwell and Company, London), although a simpler technique, using scissors and forceps, has been used successfully in the authors's laboratory, as described in *Table 1*. The washed tissue fragments are subsequently subjected to controlled enzymatic digestion in a trypsinisation vessel (*Figure 5*) to obtain a suspension of viable single cells (*Table 1*).

3.1.2 *Estimation of Cell Viability and Yield*

The final step in the pre-immobilisation treatment of the isolated cells involves a

Table 2. Determination of Cell Density and Viability.

1. Aseptically transfer equal volumes (100 – 200 µl) of the cell suspension and 0.4% (w/v) trypan blue into a small glass or polystyrene tube. Mix well, and allow to stand for 2 – 3 min.
2. Transfer a small portion to the haemocytometer chamber, and observe under the microscope. Against the pale blue background, dead cells will stain dark blue, whilst viable cells will exclude the dye and thus appear clear.
3. Count the number of viable cells on the central section of the haemocytometer grid, and express as a percentage of the total cell number present to obtain estimates of both cell yield and viability. A cellular viability of 80 – 90% should be obtained.
4. Returning to the main suspension of cells, adjust the density to that required for initial plating, using the appropriate cell culture medium.

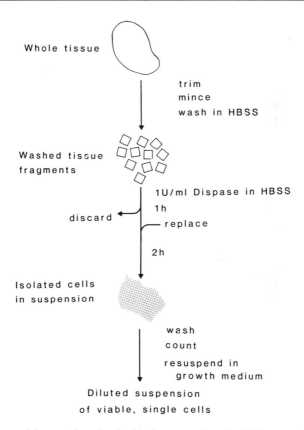

Figure 6. Summary of the procedures involved in the preparation of a viable suspension of single cells from a freshly excised tissue specimen.

determination of the cell population density, together with an estimate of the viability of the preparation. Both of these procedures are carried out with the aid of a haemocytometer, under the low-power (x 10) objective of a microscope as shown in *Table 2.*

A summary of this sequence of manipulations, progressing from whole tissue to single, viable cells, is shown in *Figure 6.*

3.2 Immobilisation of Cells as Monolayers

3.2.1 *Preparation of Replicate Cultures*

Having prepared a suspension of isolated cells from the starting tissue appropriate to the stimulator being investigated, the replicate monolayers forming the bioassay 'target' tissue should be initiated by adding successive aliquots of the starting cell suspension to the required number of bioassay wells. Since the close identity of replicate monolayers is paramount to the attainment of a high level of bioassay precision, the starting cell suspension must be distributed into the wells with the aid of a fixed-volume pipette, using disposable tips previously sterilised by autoclaving. Assuming an even distribution of cells within the suspension, and the absence of clumps of aggregated cells or tissue debris, it is possible to attain a between-culture variation in cell plating density of the order of $1-2\%$. Functionally-differentiated cells in primary culture are frequently characterised by a limited capacity for continued proliferation and should therefore be plated into each well in sufficient number to attain the desired final 'working' density at an early stage after immobilisation onto the culture support. As an approximate guide, 5×10^5 cells in a 1 ml aliquot of culture medium may be conveniently accommodated in each well of a 24-well multidish (*Figure 2*). Cells plated at higher densities may give rise to cultures having a submaximal response upon stimulation with an appropriate hormone, a feature which may be attributable to a progressive decrease in accessibilty of hormone receptors, as a result of cell crowding and overgrowth.

After initiation of the required number of replicate cultures, transfer the bioassay plates to a 37°C incubator containing an atmosphere of 5% CO_2 in air or, if medium buffered with Hepes has been used, air alone. In either case, a water-saturated incubator should be used to prevent evaporation of the culture medium.

3.2.2 *Post-immobilisation Treatment of Cells*

At intervals of $3-4$ days, remove the medium from the cell monolayers with the aid of a sterile Pasteur pipette attached to a collection vessel and vacuum line, and replace with a fresh, pre-warmed aliquot of the same medium. It is particularly important not to allow the monolayers to become dry during the medium-changing process. Cells thus immobilised in primary culture should, as a general rule, be used for bioassay purposes within $7-10$ days of culture initiation, i.e., within the period coincident with maximal culture responsiveness to hormonal challenge.

4. EXPOSURE OF IMMOBILISED CELLS TO TEST STIMULATORS

4.1 An Outline Guide to Bioassay Design

Unlike laboratory animals, tissue fragments or slices prepared from whole tissues, replicate cell monolayers prepared from a single starting tissue or cell inoculum may be considered as essentially identical with respect to their responsiveness toward a given dose of a test stimulator. Accordingly, the fundamental bioassay requirement that each individual dose of stimulator be tested on a suffi-

S.P.Bidey

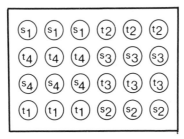

t 1-t 4 : Test stimulator doses 1 -4

s 1 -s4: Standard stimulator doses 1 - 4

Figure 7. Distribution of test and control stimulators for bioassay onto replicate cultures of immobilised cells. In this example, triplicate monolayers have been assigned to each dose of stimulator, with eight triplicate sets accommodated on a standard 24-well multidish.

cient number of test 'objects' or replicates to be representative of the whole population may, using immobilised cells, be satisfied by using only a single monolayer. It is usual, however, in the author's laboratory, for each test stimulator to be assayed on 3 − 4 replicate monolayers in order to facilitate subsequent statistical analysis of the responses obtained. The total number of cultures initiated in a given bioassay should therefore be sufficient to permit at least triplicate determinations to be made with respect to each test stimulator, together with those of an appropriate series of controls and reference standard preparations of the hormone under investigation.

Within the confines of a single bioassay plate, all incubation wells, and thus all monolayers, are subject to identical environmental influences with regard to temperature, humidity, CO_2 tension and nutrient supply. Accordingly, randomisation of bioassay test samples within a single plate is not mandatory and replicate doses of stimulator may be distributed onto adjacent cultures, as illustrated in *Figure 7*. Nevertheless, two precautions should be observed in carrying out the bioassay when using such a distribution design. Firstly, care should be taken to ensure that the time of exposure of cells to stimulator is identical for all monolayers in the plate. Thus incubations should be terminated in the same order, and over the same time-span, as stimulators were originally added to the cells. This is a particularly important consideration in the case of standard doses of stimulator, the responses to which will be used to construct a dose-response curve. Thus, if the stimulator is added to monolayers in order of increasing dose, any marked variation in incubation time between the lowest and highest doses of stimulator may introduce an element of bias into the dose-response curve. However, if the response to each hormone dose has already attained a maximal level, prior to withdrawal of the stimulator, some variation in incubation time may be tolerated without incurring a significant effect on the cellular response obtained. In order to investigate whether the responses to a series of incremental doses of a stimulator are influenced by small differences in incubation time between replicate cultures under otherwise identical conditions of incubation, any

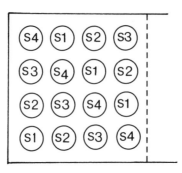

Figure 8. Distribution of four doses of stimulator within a 24-well plate, using a 4 x 4 'latin square' configuration. Each dose of stimulator has, in this example, been assayed on quadruplicate cell monolayers.

such effects may be effectively nullified by exposing cell monolayers to a series of hormone doses distributed over the bioassay plate in the form of a 'latin square' as shown in *Figure 8*, and by comparing the dose-response curve obtained with that resulting from a 'conventional' incubation scheme such as that shown in *Figure 7*.

In setting up a 'latin square' distribution, doses of hormone should be added to, and withdrawn from the cells according to the sequential order of the culture wells, e.g., horizontally, left to right, commencing with the top left-hand monolayer. Addition and withdrawal of stimulators in an identical sequence will ensure uniformity of the mean incubation time allotted in respect of each individual dose of hormone.

The second precaution that should be observed in designing and implementing a bioassay based on replicate cell monolayers relates to the importance of adopting appropriate procedures to monitor and control bioassay performance. In Section 6, the use of standardised, calibrated hormone preparations in assessing between-assay variation in cellular responsiveness is described. However, in the context of the individual bioassay using cells from a single starting tissue or subculture, quantitation of within-assay variation or 'drift' in cellular response may be made by including, in a replicate set of incubation wells on each plate in the bioassay, a known dose of reference hormone preparation. Any significant variation in the response obtained to this reference stimulator, between the replicate culture plates in a single bioassay, should be investigated and the cause rectified.

4.2 Experimental Procedures

After 7–10 days of primary culture, the immobilised cell monolayers may be used as bioassay 'target' tissue. A summary of the essential manipulations involved in carrying out a typical bioassay is shown in *Figure 9*; a detailed description of the individual stages is given in *Table 3*.

4.3 Assessment of the Bioassay Response

Upon completion of the test incubations, and extraction of the reaction product

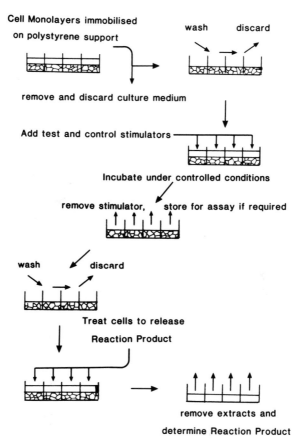

Figure 9. Sequence of cell manipulations and washing procedures involved in a typical bioassay based upon the functional response of immobilised target cells in monolayer culture.

from the immobilised cells or incubation medium, the response to each dose of stimulator is determined by quantitative estimation of the reaction product in each extract. Irrespective of the analytical technique used in such determinations, the response data obtained should be presented in the form of a calibration curve, more commonly referred to as a dose-response curve (*Figure 10*). As noted in Section 4.1, the exposure of 3 − 4 replicate monolayers of immobilised cells to each of the serial dilutions of the reference standard will enable a mean response and standard deviation (SD) to be calculated for each dilution of the stimulator. Accordingly:

(i) calculate the mean ± SD response to each dilution of the reference standard;

(ii) using linear or semi-logarithmic graph paper, plot the mean ± SD response against the appropriate dose of the reference standard stimulator; and

(iii) after constructing the dose-response curve, calculate the biologically-active dose of stimulator present in each of the test samples, by extrapolation from the calibration curve.

Table 3. Bioassay Procedure.

1. Using a sterile Pasteur pipette, remove and discard the growth medium used to maintain the cells during the initial stage of culture[a].
2. Rinse each monolayer briefly with Hank's salt solution, pre-warmed to 37°C. After washing, leave a thin film of liquid covering the monolayer, as described above.
3. Add the appropriate test or control stimulator to each monolayer, recording the sequence and time of addition.
4. Replace the cultures in the incubator, and maintain constant conditions for the duration of the incubation.
5. After the required incubation time has elapsed, remove the stimulators from the cell monolayers, using a fresh Pasteur pipette for each series of replicates[b].
6. Briefly rinse each monolayer with Hank's salt solution, as in (2) above.
7. Add 0.3 ml 12% (w/v) perchloric acid to each incubation well[c].
8. Estimate the level of the specific reaction product in the acid-soluble or insoluble phase, using a quantitative analytical technique appropriate to that material.

[a]When removing the growth medium, care should be taken to avoid contact between the pipette and the cell layer itself, since this may result in cellular damage or removal, affecting the subsequent bioassay response of the culture. It is also important that the monolayers should not be allowed to dry out after removal of the medium; a thin film of liquid sufficient to cover the cells should be retained in the wells of the bioassay plates.
[b]After the exposure of immobilised cells to test stimulators, certain reaction products may be released into the incubation medium, in levels proportional to the degree of cellular stimulation. Quantitative estimations of such products may be carried out, using standard biochemical or immunological test procedures. If, after exposure of the immobilised cells to test stimulators, the reaction product remains in the cells, an appropriate extraction technique must be applied to the cells before a quantitative estimation of the product can be undertaken. For acid-soluble products, the procedure may be continued through steps 6 − 8.
[c]The addition of acid to the cells serves a triple purpose, in that it (i) terminates the action of the stimulator; (ii) lyses the cells; and (iii) effectively separates protein and non-protein components by precipitation of the former.

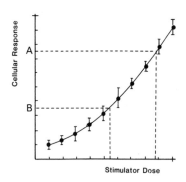

Figure 10. Construction of a dose-response curve, for which mean ± SD cellular responses are plotted against increasing doses of a standard preparation of the stimulator. Responses obtained to each test stimulator (**A, B**) are extrapolated onto the standard curve and the test doses estimated accordingly.

Figure 10 illustrates the features of a typical dose-response curve, demonstrating in particular how the biological activity of unknown test samples may be calculated from the dose-response data obtained with respect to the standard, or reference, preparation of the material. In certain cases, the limiting curvature of the typical dose-response curve may be overcome by plotting logarithms of both standard doses and responses. In such cases, the linear dose-response relationship obtained may simplify determination of the stimulator bioactivity in the test sample.

5. SPECIFIC APPLICATION: BIOASSAY OF THYROTROPHIN USING IMMOBILISED HUMAN THYROID CELLS

The principal techniques and manipulations associated with the isolation, immobilisation and maintenance of hormone-responsive cells as primary monolayers, together with the subsequent utilisation of these cultures as bioassay 'target' tissue, may be illustrated by describing studies undertaken in the author's laboratory using human thyroid follicular cells to determine the bioactivity of TSH. The objective of this investigation was to calibrate bovine TSH preparations as alternative 'secondary standards' to the less readily-available human TSH material in routine laboratory bioassays.

5.1 Cell Isolation and Immobilisation

Viable human thyroid follicular cells are obtained by enzymatic dispersion of tissue specimens removed from patients undergoing thyroidectomy. Each cell suspension prepared consists of cells derived from a single thyroid tissue

Table 4. Processing and Enzymatic Dispersion of Thyroid Tissue.

1. Treat the thyroid tissue fragments with 'Dispase II' (1.2 U/ml in Ca^{2+}- and Mg^{2+}-free Hank's salt solution), using 10 ml of enzyme solution for each gram of tissue processed.

2. After 1 h, decant the supernatant through the side-arm of the trypsinising vessel (*Figure 5*) and add an equal volume of fresh enzyme solution through the neck of the vessel, on to the tissue fragments.

3. Continue the incubation for a further 2 h, then remove the supernatant and filter through a pad of fine sterile surgical gauze into a flask containing sufficient foetal calf serum to give a final 1% serum content in the filtrate[a].

4. Upon completion of the tissue dissociation stage, harvest the isolated cells by low-speed centrifugation of the suspension at room temperature (200 g, 10 min).

5. Resuspend the pellet thus obtained with the aid of a sterile, siliconised Pasteur pipette, into 10 ml of Medium '199' (Gibco-Europe Ltd., Paisley, Scotland) containing 10% (v/v) foetal calf serum and 2 mM glutamine[b].

6. Determine the final yield and viability of cells in the suspension with the aid of a haemocytometer.

7. Adjust the density of the cell suspension, to 5 x 10^5/ml by dilution with further serum-supplemented Medium '199' containing, in addition, 100 U/ml penicillin, 100 μg/ml streptomycin and 2.5 μg/ml fungizone to counter any possible bacterial, fungal or yeast contamination of the cultures.

8. Transfer 2 ml aliquots of cell suspension into each well of the required number of 24-well polystyrene multiwell plates (*Figure 2*) (Linbro Plasticware, Flow Laboratories Ltd., Irvine, Scotland).

9. Transfer the cultures to a water-saturated incubator maintained at 37°C, under an atmosphere of 5% CO_2 in air.

[a]This two-stage enzymatic dispersion process will effectively remove, in the first stage, erythrocytes, connective tissue cells and some single thyroid follicular cells. Subsequent re-exposure of the tissue fragments to the enzyme in the second digestion stage will release cells from intact thyroid follicles, a process which may be monitored by observation of portions of the digest under low-power phase-contrast microscopy, at intervals during the dispersion process.
[b]During this procedure, any remaining enzyme activity 'carried over' with the cell pellet will be effectively inhibited by the dilution process involved in resuspending the cells.

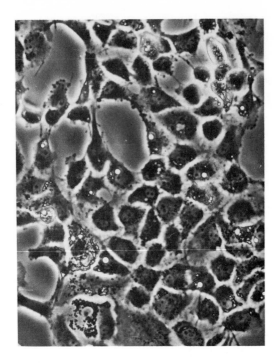

Figure 11. Human thyroid follicular cells in primary monolayer culture, observed 96 h after plating, by phase contrast microscopy (x 460).

specimen only. After washing the excised tissue in Hank's salt solution, the initial processing stages may be carried out according to the procedure described in Section 3.1 and is summarised in *Table 4*.

Four days after the initiation of thyroid follicular cells in culture, individual cells will have become flattened, and immobilised on the base of the polystyrene culture plate (*Figure 11*). These monolayer cell cultures may now be used as TSH bioassay target tissue. However, if the medium is renewed at 3 – 4 day intervals, the monolayers will remain responsive for a further 4 – 7 days before dedifferentiation becomes apparent and fibroblast overgrowth predominant.

5.2 Bioassay of Thyrotrophin (TSH)

Several preparations of bovine TSH, obtained from various commercial sources, have been established as 'secondary' reference standards for the hormone in the author's laboratory, after calibration against human TSH. The biological activity of all TSH preparations tested has been determined on the basis of the stimulation of intracellular cAMP accumulation observed in primary monolayer cultures of human thyroid follicular cells, as determined after 4 – 11 days of *in vitro* maintenance. The experimental procedures followed in these calibration studies are described in *Tables 5* and *6*, with reference to a single preparation of bovine TSH ('Thytropar'; Armour Pharmaceuticals, Kankakee, Illinois, USA) supplied in lyophilised form with a lactose carrier, and with a stated 'activity' of 10 units

Table 5. Procedure for the Bioassay of Thyrotrophin (TSH).

1.	Reconstitute both the reference standard and unstandardised preparations of TSH from the lyophilised form to give appropriate initial dilutions in Hank's salt solution.
2.	Prepare a range of serial dilutions of each TSH preparation[a].
3.	To initiate the bioassay of TSH, remove the growth medium from the cultures by aspiration, and wash the cell monolayers once with Hank's salt solution. Then add to each monolayer the appropriate dose of TSH, prepared in Hank's salt solution as described above, together with 0.5 mM 3-isobutyl-1-methylxanthine (MIX)[b].
4.	Incubate all cultures at 37°C for 20 min under an atmosphere of 5% CO_2 in air.

[a]In such a series of dilutions, the lowest dose of hormone tested might be 1/100 of that of the most potent hormone dose. Prepare a sufficient volume of each working hormone dilution to give 3 x 1 ml replicates (i.e., 1 ml for each of the three replicate monolayers of bioassay 'target' cells used for each test determination). Eight dilutions of each TSH preparation will generally be sufficient to establish the potency of the unknown preparation relative to that of the 'primary standard'.

[b]Inclusion of MIX within the incubation medium effectively inhibits the activity of intracellular phosphodiesterases, giving an enhanced cAMP response to each TSH test dose by inhibiting degradation of the nucleotide.

Table 6. Cyclic AMP Extraction and Assay.

1.	After completion of the incubation period with TSH, rapidly aspirate the incubation medium from the cell monolayers.
2.	Add 0.5 ml of ice-cold absolute ethanol to each incubation well.
3.	Seal the culture plates and transfer to a −20°C freezer for 24 h[a].
4.	Remove precipitated protein by centrifugation (2000 g, 15 min, 4°C).
5.	Transfer 200 μl portions of the ethanolic fractions to small glass test tubes, and evaporate to dryness under a stream of nitrogen.
6.	Reconstitute the dried residues, containing cAMP, in 25 mM Tris/50 mM NaCl/8 mM theophylline/6 mM mercaptoethanol ('cyclic AMP assay buffer').
7.	Determine the cAMP content of the cell extracts.
8.	Express the final cAMP levels attained within each set of triplicate cultures, in response to each TSH dose, as picomoles per culture (mean ± SD)[b].

[a]This treatment effectively lyses the cells and precipitates intracellular proteins, whilst cAMP remains in solution. If desired, the cells may be scraped into the ethanol with the aid of a rubber policeman to effect complete homogenisation of cell components.

[b]This calculation assumes that the cell population densities of replicate cultures are identical, which may be tested by determining the protein or DNA contents of a series of representative, replicate monolayers. For preparations derived from a single plating suspension of cells, the between-culture variation in DNA content should approximate to ± 5% if the initial plating of cultures was carefully performed.

per ampoule. The highly-purified preparation of human TSH, against which the bovine TSH is calibrated as a 'secondary standard', is the First International Reference Preparation of human thyroid-stimulating hormone (Pituitary TSH), coded 68/38, with an assigned activity of 150 mU/ampoule.

In the author's laboratory, cAMP has been determined by a competitive protein-binding saturation assay technique (14) using a naturally-occurring binding protein with a high specificity for the nucleotide. However, a variety of alternative procedures (15) may be substituted, including that based upon a specific cAMP antibody (16). The principal advantage to be gained by adopting the latter technique is an increased assay sensitivity, which enables the determination of

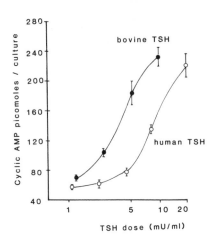

Figure 12. Accumulation of cAMP in primary monolayer cultures of human thyroid cells, in response to incubation with bovine TSH ('Thytropar', Armour Pharmaceuticals) or human TSH (First International Reference Preparation, coded 68/38). Bovine TSH activity is expressed in units assigned to the preparation by the manufacturer, and human TSH activity is expressed in standard units assigned by the National Institute of Biological Standards and Control, London. Incubation time was 20 min in both cases, and results plotted are the mean ± SD response of triplicate cultures exposed to each test dose of TSH.

very low cAMP levels. However, for most applications, if sufficient amounts of cAMP are available for assay, the competitive protein-binding assay is both convenient and reproducible. The binding protein itself is stable for long periods of storage at −20°C (2−3 years) and the tritiated cAMP is readily-available from commercial sources.

5.3 Evaluation of Dose-response Data

(i)   Construct dose-response curves for the intracellular cAMP levels attained in thyroid cell monolayers in response to bovine and human TSH respectively, as illustrated in *Figure 12*.

A given relationship between stimulator dose and cellular response will only hold true for a particular set of assay incubation conditions and, in many cases, will be critically dependent upon the bioassay incubation time, together with the temperature, pH and ionic strength of the incubation medium. In comparative studies such as this, therefore, it is of particular importance that identical bioassay incubation conditions should be maintained with respct to the two TSH preparations under study. It should also be noted that the dose-response curves, as plotted in *Figure 12*, are not linear, and a more convenient analysis and comparison of the two sets of data may be made after logarithmic transformation of both dose and response data. The linear plots thus obtained from *Figure 12* are shown in *Figure 13*. It can be established, by appropriate statistical analysis of the data (17), that the dose-response curve for human TSH is not significantly non-parallel with that obtained with respect to bovine TSH, with the slopes, in this ex-

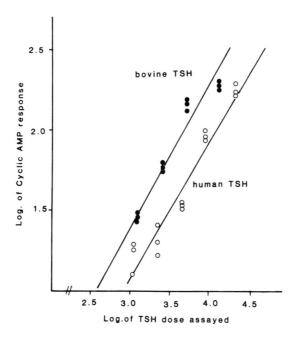

Figure 13. Logarithmic transformation of the dose-response data obtained in respect of bovine and human TSH as shown in *Figure 12*. Individual responses are shown for each of the three replicate monolayers exposed to each TSH dose.

ample, being 0.89 and 0.91 for the human and bovine TSH dose-response curves, respectively.

(ii)   If a constant relationship between the cellular responses to increasing bovine and human TSH doses can be demonstrated, as illustrated in *Figure 13*, estimate the potency ratio of the two TSH types.

In the example shown, the potency of the bovine TSH preparations is calculated to be 2.5 times that of the human TSH. Accordingly, each dose of bovine TSH assayed may be recalibrated in terms of human TSH units, and the former material effectively substituted for the less-readily available human TSH as a convenient 'secondary standard' in the laboratory bioassay for human TSH.

6. THE MONITORING AND CONTROL OF BIOASSAY PERFORMANCE

Essential to the long-term objective assessment of bioassay performance, the implementation of appropriate, well designed quality-control procedures is, nevertheless, a frequently overlooked aspect of bioassay design. Such procedures, which essentially involve a quantitative assessment of between-assay variation in response to fixed doses of a series of standardised reference materials, assume particular importance in the case of bioassay systems based upon cells in primary monolayer culture, the responses of which frequently reflect the variability of different starting tissue preparations. Accordingly, the principles underlying the design of such a control system will be briefly discussed in this section.

6.1 The Importance of Bioassay Control

One of the principal characteristics of hormonally-responsive, immobilised cells in primary culture is an inherent variability in responsiveness of such preparations to *in vitro* stimulation. Accordingly, the ongoing performance of any series of bioassays based upon the responses of cells prepared from different starting tissues should never be assumed to be constant, and even for those bioassays based upon the responses of continuous cell strains or established cell lines, unintentional and random variations in growth medium components, or between different batches of the serum used as a medium supplement, may give rise to gradual alterations in the growth pattern or response characteristics of the 'target' cells. Overall culture responsiveness may also be affected, in the case of primary cultures, by the variable presence of non-responsive, or different 'contaminating' cell types, whilst in the case of continuous, or established cultures, one should always consider the presence of mycoplasma as a possible underlying cause of altered cellular response characteristics.

6.2 A Simple Bioassay Monitoring Scheme

Any procedure designed to monitor the performance of a sequential series of bioassays, either as a check of the responsiveness of individual batches of cells derived from different starting tissues, or as a means of detecting long-term alterations in response characteristics of a single, continuous cell strain or established line, should involve a repeated determination of the cellular response to a fixed dose of a reference or 'control' stimulator or a series of such reagents, each with an established 'potency'. In such a procedure, any marked deviation of the observed cellular response from a pre-defined 'mean' value, established from the responses attained in previous bioassays, may be indicative of an alteration in culture characteristics such that rejection of that particular set of bioassay data

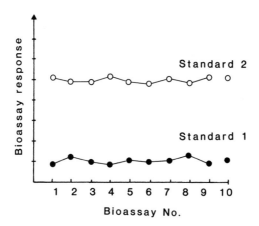

Figure 14. A simple quality-control scheme to monitor bioassay performance, based on the responses obtained in respect of two standard preparations of stimulator (1, 2) in each of a series of sequential bioassays. For each standard tested, the response obtained may be compared with the mean ± SD response obtained in the preceding bioassays in the sequence.

may be indicated. In practice, for the majority of bioassay applications, the basic requirements of this simple monitoring procedure will have been satisfied by inclusion in the bioassay of appropriate doses of the reference standard of the hormone under investigation, as used to obtain the dose-response curve. However, in more comprehensive performance-monitoring schemes, the chosen hormone doses should include several from within the range most likely to be encountered in the test samples. In the case of serum hormones in man, this value will be within the accepted normal physiological range of the hormone. Additional doses of reference standard may also be included, to correspond for example with the limit of acceptable assay sensitivity (i.e., very low levels of hormone), whilst doses in excess of the normal serum range of the hormone may be required if the bioassay is being used to monitor raised hormone levels in serum. For each sequential bioassay, which may use cells derived from different starting tissues, or subcultures of an established cell line, the responses obtained with respect to each of the individual reference doses may be plotted as shown in *Figure 14*. In this manner, the response obtained to a given dose of hormone, in a single bioassay, may be compared with the mean response observed with respect to the same dose of hormone tested in the series of bioassays immediately preceding that being monitored. Bioassays in which the responses fall outside of the mean ± SD range established for this preceding group may then be regarded as anomalous and rejected.

7. RECOMMENDED SUPPLIERS OF CELL CULTURE REAGENTS AND MATERIALS

7.1 Culture Apparatus (Glass)

Arnold R.Horwell Ltd.,
2 Grangeway,
Kilburn High Road,
London NW6 2BP, UK

Bellco Glass Inc.,
Vineland,
NJ 08360, USA

7.2 Disposable Plasticware and Culture Reagents

Flow Laboratories Ltd.,
PO Box 17,
Second Avenue Industrial Estate,
Irvine KA12 8NB, UK

Flow Laboratories Inc.,
7655 Old Springhouse Road,
McLean,
VA 22102, USA

Gibco Europe Ltd.,
Trident House,
PO Box 35,
Renfrew Road,
Paisley PA3 4EF, UK

Grand Island Biological Co.,
Grand Island,
NY 14072, USA

8. REFERENCES

1. Ekins,R.P. (1978) in *Radioimmunoassay and Related Procedures in Medicine*, Vol. 1, International Atomic Energy Agency, Vienna, p. 241.

2. Grinell,E. (1978) *Int. Rev. Cytol.,* **53**, 65.
3. Ambesi-Impiombato,F.S., Parks,L.A.M. and Coon,H.G. (1980) *Proc. Natl. Acad. Sci. USA,* **77**, 3455.
4. Ambesi-Impiombato,F.S., Picone,R. and Tramontano,D. (1982) in *Growth of Cells in Hormonally-defined Media, Cold Spring Harbor Conference on Cell Proliferation,* Vol. **9**, Sato,G.H., Pardee,A. and Sirbasku,D.A. (eds.), New York, p. 483.
5. Bottenstein,J., Hayashi,I., Hutchings,H., Masui,J., Mather,D.B., McClure,S., Ohasa,A., Rizzino,A., Sato,G., Serrero,G., Wolfe,R. and Wu,R. (1979) *Methods Enzymol.,* **58**, 94.
6. Barnes,D. and Sato,G. (1980) *Cell,* **22**, 649.
7. Bidey,S.P., Marshall,N.J. and Ekins,R.P. (1981) *Acta Endocrinol.,* **98**, 370.
8. Powell-Jones,C.H.J., Thomas,C.-G. and Nayfeh,S.N. (1980) *J. Biol. Chem.,* **255**, 4001.
9. Ekins,R.P. (1974) *Br. Med. Bull.,* **30**, 1.
10. Bidey,S.P., Marshall,N.J. and Ekins,R.P. (1981) *J. Clin. Endocrinol. Metab.,* **53**, 246.
11. Bidey,S.P., Marshall,N.J. and Ekins,R.P. (1982) *Acta Endocrinol.,* **101**, 359.
12. Kruse,P.F. and Peterson,M.K. (1973) *Tissue Culture Methods and Applications,* published by Academic Press, New York.
13. Paul,J. (1979) *Cell and Tissue Culture,* Fifth Edition, published by E. and S. Livingstone, London.
14. Brown,B.L., Albano,J.D.M., Ekins,R.P., Sgherzi,M.A. and Tampion,W. (1971) *Biochem. J.,* **121**, 561.
15. Holmegaard,S.N. (1982) *Acta Endocrinol.,* **101**, Suppl. 249.
16. Steiner,A.L., Wehmann,R.E., Parker,C.W. and Kipnis,D.M. (1972) in *Advances in Cyclic Nucleotide Research,* Vol. **2**, Greengard,P. and Robison,G.A. (eds.), Raven Press, New York, p. 51.
17. Finney,D.J. (1978) *Statistical Methods in Biological Assay,* edn. 3, published by Griffin, London.

Index

INDEX

173

Forthcoming

Nucleic acid hybridisation
a practical approach
Edited by B D Hames and S J Higgins
A practical laboratory-bench manual of techniques for identifying and analysing the structure of specific gene sequences. This book is unique in bringing together the techniques' major applications at both the theoretical and the practical levels.
Due September 1985; 250pp (approx); 0 947946 23 3 (softbound)

Animal cell culture
a practical approach
Edited by R I Freshney
After an introductory chapter dealing with basic techniques, this book provides detailed protocols both for traditional areas like organ culture, characterisation and storage, and for those in developing fields. These include serum-free media, cell separation and *in situ* hybridisation.
Due October 1985; 250pp (approx); 0 947946 33 0 (softbound)

Photosynthetic energy transduction
a practical approach
Edited by M F Hipkins and N R Baker
An up-to-date laboratory manual for researchers and students wishing to learn a wide range of techniques for the study of photosynthetic energy transduction.
Due late 1985; 250pp (approx); 0 947946 51 9 (softbound)

Biochemical toxicology
a practical approach
Edited by K Snell and B Mullock
Chapters written by laboratory experts provide practical guidance and 'tricks of the trade' for the most useful techniques in toxicological research. The book is unique as a guide for researchers at all levels, especially those in pharmaceutical and agrochemical laboratories.
Due late 1985; 250pp (approx); 0 947946 52 7 (softbound)

PRICES TO BE ANNOUNCED

◇ IRL PRESS

IRL Press Ltd, PO Box 1, Eynsham, Oxford OX8 1JJ, UK
IRL Press Inc, Suite 907, 1911 Jefferson Davis Highway, Arlington, VA 22202, USA